Nature's Balancing Act

Nature's Balancing Act

*How Small Quirks of Physics Makes
Life Possible*

Ken Hicks

*Professor, Department of Physics and Astronomy,
Ohio University, United States of America*

OXFORD
UNIVERSITY PRESS

Oxford University Press is a department of the University of Oxford.
It furthers the University's objective of excellence in research, scholarship,
and education by publishing worldwide. Oxford is a registered trade mark of
Oxford University Press in the UK and in certain other countries.

Published in the United States of America by Oxford University Press
198 Madison Avenue, New York, NY 10016, United States of America.

CIP data is on file at the Library of Congress

ISBN 9780197771440

DOI: 10.1093/9780197771471.001.0001

Printed by Integrated Books International, United States of America

The manufacturer's authorised representative in the EU for product safety is Oxford University Press España S.A. of El Parque Empresarial
San Fernando de Henares, Avenida de Castilla, 2 – 28830 Madrid (www.oup.es/en or product.safety@oup.com). OUP España S.A. also acts as
importer into Spain of products made by the manufacturer.

Contents

Preface

Yogi Berra once quipped, "If the world were perfect, it wouldn't be." Presumably he was talking about baseball and that imperfections in play are what make the game fascinating. His words could just as easily be applied to the universe; for if the universe were perfectly symmetric, then we wouldn't be here. Indeed, that saying by Yogi Berra perfectly captures the idea behind this book: "Why are we here?"

Nature has an exquisite balance of physical laws that makes it possible for life (as we know it) to exist. There are no fewer than ten ways that, if the balance of physical laws were changed ever so slightly, the resulting universe would be unrecognizable from our own. Perhaps some other form of life could exist in that hypothetical universe, but certainly life as it exists on Earth could not be present there.

While some people might consider that exploration into "how the universe could be different" as a purely academic question, deeper questions arise as we explore this topic. Why is Nature so perfectly balanced? Are there other physical laws, yet to be understood, that cause the universe to be this way? Is there a Grand Unified Theory that could explain everything in Nature? So, this journey into the unknown is worth taking. It brings awareness that the universe we live in is delicate in its construction and that life is precious.

Chapter 1 starts with a fundamental question: "How can the entire universe appear out of nothing?" The answer to this is often sidestepped; our current understanding of physical laws cannot tell us what was present before the Big Bang. However, there is more to this story. In recent decades, the theoretical concept of "inflation" (named in part because it ignited the Big Bang) has grown in popularity. One of the goals of this book is to explain inflation in language that is easy to understand and hence explain how known physical laws allow something (the universe) to appear out of nothing.

But the main topic of this book goes well beyond the Big Bang. Later chapters tackle other fundamental questions, such as "Why are we here?" and "How is the universe structured to allow life to exist?" The former question is perhaps more relevant to the field of philosophy than it is to science. But it also serves as a guide to formulating more specific scientific questions, such as "How do complex molecules form?" and "What physical laws are needed for life to begin?"

In exploring the answers to such scientific questions, it becomes evident that Nature is slightly off-balance, in just the way needed for life (as we know it) to exist. One could even say that the universe is perfectly situated for life to exist. Of course, you can turn that around and say that life exists simply because the universe formed this way, and hence the formation of life is inevitable. This is the anthropic principle. If the universe had slightly different physical laws, then there might be no humans to ask the question of why life exists!

I find the anthropic principle to be fascinating and a source of endless discussions. Some people like to take a philosophic approach and argue about why the universe must be the way it is. This suggests that physical laws would force any universe to be similar to our own. Others might take a religious point of view, arguing that there must be an intelligent being that designed the universe to be just so. But the most popular discussions speculate about the existence of a "multiverse," a larger group of universes, each with its own physical laws, of which our universe is just one part. While there is no evidence of the multiverse (either for or against this hypothesis—since we can't see outside of our local universe), that doesn't stop theorists from creating mathematical models of a multiverse (or movies being made based on this idea). The question is far from settled.

Whether we live in a universe or a multiverse, our home has physical laws that are nicely balanced (or in some cases, slightly off-balance) so as to allow life to exist. Exploring the balance of Nature is the primary goal of this book.

In Chapter 3, we explore the balance of matter and antimatter in Nature. Astronomical observations show that our universe appears to be dominated by matter, with only a tiny amount of antimatter. This tells us that there must be a slight imbalance in the physical laws for creating matter and antimatter, which so far has eluded experimental physics. Yet this imbalance is necessary for life to exist. Without it, our universe would have ended eons ago, with equal amounts of matter and antimatter annihilating each other.

Of course, the balance of matter and antimatter is not the only aspect of Nature that is important for life to exist. There are other delicate balances that are not so obvious. For example, what if the strong nuclear force were a bit stronger? By "strong force," I mean that it takes a lot of force to pull apart the protons and neutrons in the atomic nucleus. If the nuclear force were even stronger, it would take more energy to separate a proton (or a neutron) from its parent nucleus. At first, you might think, "Who cares?" Upon delving deeper, though, it becomes clear that life could not exist if the nuclear force differed by just a few percent!

In addition to the main text, each chapter has text in boxed areas with related topics that will interest the reader. For readers who want a compact story, the main text can be read without referencing the boxed text. A richer story develops with the boxed text that provides additional context.

For full disclosure, almost all the ideas presented in this book were first developed by other scientists. My role is more akin to that of a journalist, to report on these scientific advances (hopefully in a way that the average reader can understand). For many years, I wrote a popular science column for my local newspaper, *The Columbus Dispatch*. My background is in the field of nuclear and particle physics, for which my career, with hundreds of publications in peer-reviewed scientific journals, stretches back over forty years. It is in this context that I've read scientific papers and discussed with my colleagues the ideas that are presented herein. My contribution is to synthesize, in one book, the remarkable balances (and imbalances) found in Nature that make human life possible. Of course, any mistakes in the presentation are mine alone.

This book is intended for a person with some background in science, at the level taught in high school. While a complete understanding of the scientific concepts requires complex mathematical equations, none of these appear here. A deeper appreciation of the topics discussed here can be had by taking upper-level college physics and math courses or by reading more advanced books on the subject. Put simply, I cannot do justice to the topics here by using words alone. It is my hope that readers will use this book as a starting point to learning more about physics and cosmology.

I would like to acknowledge the many colleagues who contributed to this book, either by informal discussions or reading excerpts of chapters. I want to thank faculty colleagues who read earlier drafts of chapters, including Professors Clowe, Hines, Horowitz, Smith, Tees, and Ulloa, among others. I am especially grateful to Gillian Berchowitz, formerly of Ohio University Press, who encouraged me at a critical point in the writing process, which led me to contact the editors at Oxford University Press (OUP). The team at OUP has been wonderful to work with. Finally, I want to thank my wife for her advice and encouragement. I am forever grateful for her loving support.

Chapter 1
Explosive Ideas

Nothing will come of nothing . . .

—King Lear, Act I Scene I

Common sense is an amazing thing when you think about it. From everyday experiences, we notice patterns around us. Those patterns feed into our unconscious, helping us to make sense of our world. For example, common sense tells us that objects don't appear out of thin air, like magic. So how did the Big Bang happen? How did the entire universe appear out of nothing?

If you go to a magic show, you might see an act that defies our common sense, like making coins seemingly appear out of nothing. Some acts even have elephants that vanish from the stage, just to appear moments later. It's fun to see these performances, but down deep we know that there's some kind of trick: sleight of hand or smoke and mirrors. Our common sense tells us that in real life things don't appear out of nowhere.

Similarly, some laws of science may seem to go against our common sense. This can be something as simple as growing plants. Consider the simple case of a small tree growing in a pot outdoors. Suppose that the pot keeps the tree and soil isolated, except for exposure to the outside weather. If you weigh the pot each month, the tree gets heavier. Where did the extra mass come from? Water was added, but the tree isn't made just of water. Could its mass be increasing out of nothing? This was a hotly debated topic hundreds of years ago, along with questions about whether life could occur by spontaneous generation. The latter hypothesis wasn't disproved until the mid-nineteenth century (Pasteur, 1882).

In modern times, we know, of course, that the plant absorbs water and, combined with carbon dioxide from the air, makes hydrocarbons that increase the tree's mass. It really did come out of thin air, and it couldn't grow with no air. We're taught that mass is always conserved in chemical reactions, as long as we account for all of the sources. This new concept, that mass is conserved, gets integrated into our common sense. But as we shall see, even the conservation of mass is not always true, as Einstein first discovered. Mass can be converted to energy. This violation of our modern common sense is not

Nature's Balancing Act. Ken Hicks, Oxford University Press. © Oxford University Press (2025).
DOI: 10.1093/9780197771471.003.0001

as easy to observe in daily life. (It only happens in situations such as nuclear explosions or the nuclear reactions that occur in the sun.)

For another example of how science appears to go against common sense, a principle of physics states that energy (in all of its forms) is conserved. For simplicity, let's consider just one familiar form of energy, called kinetic energy, which is the energy of a body in motion. Isaac Newton, the founder of the field of physics, wrote in his master treatise, the *Principia*, his First Law, paraphrased as, "A body in motion will stay in motion unless acted on by an outside force." This flies in the face of what Aristotle wrote, about 2,000 years before Newton, that moving bodies always come to rest. Aristotle was simply observing the world around him, and if a stone rolls down the hill, it eventually stops. To him, it made sense that kinetic energy disappeared. He didn't have modern tools, such as a wheel with nearly frictionless bearings, to show that objects will continue to move if there's little or no friction. It wasn't until the mid-eighteenth century that the conservation of energy was established (Whitaker, 1975). In other words, the concept of energy (and its conservation) was not easily established based on everyday observation. Instead, we must rely on careful measurements to aide in our understanding of Nature.

Taking this idea a bit further, energy is conserved if we consider *all forms of energy*. That includes the energy it takes to heat up an object. Friction will transfer kinetic energy into heat energy. An easy way to visualize this is to think about stopping a moving car. If you were to feel the brakes after a moving car stops, the brakes would be warm to the touch. Going the other way, a car is propelled by burning fuel, and the heat energy in the engine is converted into kinetic energy of the car. This balance between kinetic energy and heat energy was first formulated mathematically by the French physicist Joule (Joule, 1845) and is now known as the first law of thermodynamics.

But is our information complete? Are there other forms of energy that need to be accounted for? You won't be surprised to hear that the answer is yes. For example, there is chemical energy, where energy is stored in the bonds between atoms, and this energy can be released under the right circumstances, such as giving gasoline a spark or detonating a stick of TNT.

Another form of energy is mass energy, which wasn't understood until Einstein came up with his famous equation, $E=mc^2$. This equation equates energy with mass (times a constant, c, which is the speed of light). It is this principle of the theory of **special relativity** that allows a nuclear bomb to explode, turning some of the excess mass in the uranium nucleus into energy. The principle is simple: energy is conserved if we account for all forms of it.

You might be thinking, "But what does all this have to do with the Big Bang?" or "Doesn't the Big Bang violate the conservation of energy?" How did all the mass of the universe, along with the billions and billions of stars, come out of nothing? At the face of it, the Big Bang seems like a ridiculous notion. Even if you allow that mass can come from energy, where did all that energy come from? In essence, what went bang? Keep reading if you want to know.

The short answer is that we can't yet prove what started the Big Bang. However, there is a hypothesis called **inflation** that provides a reasonable answer to account for the conservation of energy. Scientists don't yet have firm evidence for this hypothesis, but a lot of anecdotal evidence supports it—enough that it has become the leading idea for what set off the Big Bang.

To give an analogy, a police detective might know that someone was murdered, along with a general idea of how the person was shot, but not know who pulled the trigger. In the case of the Big Bang, the "who pulled the trigger" can be translated as "what subatomic particle sparked the explosion." Similarly, the "how was the person shot" becomes "how do we account for all forms of energy." For the Big Bang, we have a general idea of "how," but not any evidence for the "who."

To understand what follows, let's relinquish some closely held principles and entertain some new ideas. As mentioned above, Newton had to relinquish the teachings of Aristotle to formulate his first law of motion. Similarly, Einstein had to relinquish the old concept that energy and mass are separate and incontrovertible to come up with his equation of $E=mc^2$. Now, we need to move beyond Einstein's special theory of relativity ($E=mc^2$) to include his **general theory of relativity**. The general theory, which Einstein published (Einstein, 1915) about ten years after formulating his special theory, includes a balance of mass, energy, and gravitation. So, an equation that equates just mass and energy, like in $E=mc^2$, is not sufficient when the gravity is sufficiently strong (like near a black hole). Another form of energy (gravitational energy) must also be considered. In mathematical language, we say that the stress-energy tensor[1] is balanced with the curvature of space-time (Misner, Thorne, and Wheeler, 1973).

Put more simply, we need to allow for the fact that space itself (or more precisely a combination of space and time called **space-time**) can stretch or compress. Our common sense (with help from geometry) tells us that space is immutable, so that all measurements of space can be done with a ruler of fixed length. But Einstein discovered that stretching (or curving) of space-time is

---

[1] A tensor is a math term, similar to a vector, but with two indices.

the same as a gravitational field (see Box 1.1). Einstein's equations of general relativity have been used to make a variety of precise predictions, all of which have been verified by experimental measurements (Will, 2014).

For example, common sense would tell us that an atomic clock (one of the most precise measures of time we have) in the basement of a building would keep time at the same rate as an atomic clock on the top floor. This is not the case, as shown by precise measurements (Pound and Rebka, 1981). This happens because the gravitational field is slightly smaller as you move away from Earth. Similarly, common sense tells us that light travels in a straight line, like in a laser alignment tool. But Einstein predicted that light would bend in a strong gravitational field, and astronomical measurements showed that he was right (Eddington, 1920). There are many measurements that could be described, but the bottom line is that space-time can be distorted, and it takes a lot of energy (or the presence of a large mass) to do this.

Box 1.1 The Gravitational Field

A basic concept in physics is the topic of energy, which is seen in various forms. For example, there is *kinetic energy* (the energy it takes to move a massive object, like a car moving on a highway) and *potential energy* (the energy tied up but available to impart to an object, like a tightly coiled spring). The concept of the gravitational field was introduced to explain how a ball gains energy when dropped. In the simplest terms, the gravitational potential energy (held by this invisible "gravitational field") is transferred to the ball, giving it kinetic energy as it drops.

But does the gravitational field really exist? In truth, the gravitational field is an imaginary concept that is used to simplify the physics calculations. Similarly, the concept of an electric field is used to explain how charged objects (like an electron) move when placed near an opposite charge (and the magnetic field is similarly used for magnets). The concept of a "field" is useful in high-school physics, but in reality Nature is more subtle than this.

It was Einstein who first discovered that the field equations for gravity are the same as those used by mathematicians for tracking particles on a curved surface. From this, it was shown that gravity causes space to "curve."

Think of empty space as being filled with imaginary boxes, sort of like graph paper that has a grid printed on it. When the paper is flat (or the space is empty), each box represents one unit of distance. Now scrunch up the paper (or introduce masses into empty space) and the lines no longer go straight.

Perhaps a better analogy would be to think of a grid printed on a balloon and stretching the surface (or blowing up the balloon). The distances of grid points are no longer the same all over the surface. So, in analogy, think of space itself as squishable, so that it can be stretched or compressed. It takes energy to stretch space, just like it takes energy to stretch a balloon. Once stretched, it can give back energy when released.

So, think of the gravitational "field" as representing the stretching of space. When you release a ball, it gains energy because the curvature of space has changed ever so slightly. In a physics class, we might say it gains energy from the gravitational field, but really it's the same thing.

Why, then, don't we notice the stretching of space? The answer is because it takes an enormous amount of energy to make this noticeable on the human scale. In order bend light from a star, as first predicted by Einstein (and measured in 1919 by Eddington), it takes the mass of the sun (or more) to make even a tiny deflection. The change in space curvature from dropping a ball is invisible to the human eye (and too small to measure with any instrument).

The takeaway message is that energy can be stored (or retrieved) by changing the curvature of space itself. This may seem to be a radical concept (as compared with the comfortable guise of the gravitational field), but it comes directly from Einstein's equations.

A good way to demonstrate this principle of space-time is the announcement in 2015 by the **LIGO** Collaboration (see Box 1.2) that they measured gravitational waves that moved across Earth, which originated when two black holes merged together in a far-off galaxy (Abbot et al., 2016). Gravitational waves are a stretching of space-time that propagate outward from a disturbance (the black holes colliding), much the same as water waves propagate out in a circle when the surface of water is disturbed (such as throwing a stone into the water). Gravitational waves can't be seen or heard by the human senses, because the stretching of space-time is too small. But very sensitive devices like LIGO can now measure this stretching using laser light. The technology that goes into LIGO is fascinating (and could take a whole book to describe). The important point is that the measurements fit very precisely to the wave shape predicted by Einstein's equations for the case of two black holes orbiting each other, losing energy by radiating gravitational waves, and then eventually (as the orbit gets smaller and smaller) merging into one, bigger, black hole.

Box 1.2 The LIGO Facility

The Laser Interferometer Gravitational-wave Observatory (LIGO) was built with the goal of detecting a new phenomenon: a gravitational wave. These waves are invisible to human perception but measurable with very sensitive devices. Think of it as a super-sensitive seismometer. You may not be able to feel the ground shaking for a small earthquake that is hundreds of miles away, but a good seismometer can pick it up. Similarly, LIGO is designed to feel the shaking of space itself due to the passage of a gravitational wave. These waves are caused by cataclysmic astronomical events such as two black holes colliding.

The idea of gravitational waves goes back to a time before Einstein, but it wasn't until the theory of general relativity that a mathematical framework was available to calculate them. Einstein had initially had doubts about whether gravitational waves were a result of general relativity. Due to the approximations used in the original solutions, though, he eventually was convinced that they were real. However, he also believed that the effects of gravitational waves were too small to ever be detected. Of course, even Einstein couldn't predict the incredible pace of technology that has made LIGO possible.

LIGO is based on the technology of detecting light waves. When a beam of light is split apart—using, say, a partially silvered mirror (where some of the light is reflected and some passes straight through)—and then brought back together, the light interferes with itself to create a series of light and dark bands. Think of water waves from a passing speedboat that reflect off a retaining wall going back to hit the incoming waves. The ripples seen are the same effect as the interference pattern from the split beam of light.

If the space is stretched along one of the arms of the split light but not the other, then the interference pattern will change. The challenge here is that the gravitational wave is weakened by the time it reaches Earth, and the stretching of space might be only a thousandth of the size of a proton! This is so small that it's easy to see why Einstein said that it would never be detected. But the genius of the scientists who built LIGO found a way to measure this tiny stretching and contracting of space as the gravitational wave passes by.

The first gravitational waves were detected in 2015 from the merging of two black holes in a far-off galaxy about 1.3 billion light-years away. Other detections followed, including a gravitational wave made from the merging of two neutron stars (more on these later). For this, the 2017 Nobel Prize in Physics was awarded to three Americans: Rainer Weiss, Kip Thorne, and Barry Barish. Of course, a large team was involved in the conception, design, and building of the two LIGO facilities.

In 2017, I was honored to be asked (by the National Science Foundation) to chair a review panel on future funding for the LIGO facility. This was a fascinating review, considering all the breakthrough results that had just been achieved. The future of gravitational wave detection is bright, and there are still many ideas about how this technology could be improved. Now astronomers have a new way to probe the cosmos in addition to telescope observations.

What is interesting about the merger of two black holes is that the Einstein equations allow us to deduce, from the LIGO measurements, the masses of the original two black holes, as well as the mass of the final (merged) black hole. The mass of a black hole is proportional to the amount of material that has fallen into it. For example, if a black hole swallowed up our sun, its mass would increase by "1 solar mass." Because the mass of our sun is a convenient "unit," the masses of black holes are often quoted in terms of solar masses. The first detection of gravitational waves (Abbot et al., 2016) was due to the merger of two black holes having masses of about 35 and 30 solar masses, respectively. (I say "about" because there is some measurement uncertainty in these masses.) The final black hole has a mass of only about 62 solar masses. Where did the extra mass go? It didn't disappear. Recall that mass can be turned into energy, and energy can be used to stretch space-time. In other words, the mass went into making the gravitational waves, just as Einstein's theory of general relativity predicted.

Before going on, let's recap. First, energy comes in many forms, and over 200 years ago it was established that energy is conserved, meaning that energy doesn't just disappear (or appear), but shifts from one form to another. This includes shifting from, say, kinetic energy to heat or to mass. Einstein's general theory of relativity extends this principle to include energy in the form of stretched space-time. Einstein also showed that curved space-time is proportional to the gravitational field, and the more curved (stretched) it is, the stronger the gravity. This concept becomes very important in the next section, where the initial stage of the Big Bang is discussed.

1.1 Theory of Inflation

The inflation scenario was first proposed by Alan Guth (see Box 1.3) in the early 1980s (Guth, 1981) and then further developed by others, see (Linde, 1982) and (Albrecht and Steinhardt, 1982). In one variant of the inflation scenario,[2] the universe started from a quantum fluctuation (Tryon, 1973), producing an unstable particle (or several particles) that sparked the Big Bang. There's a lot there in that sentence, so let's pull it apart and discuss each piece.

Box 1.3 Alan Guth

Few people have made such an impact on cosmology as Alan Guth, who was the first to publish the theory of inflation. Experimental proof of this theory is not yet complete, but most cosmologists believe that something like inflation must have happened at the earliest stages of the Big Bang. This theory can answer many questions that arise about what started the Big Bang. If inflation theory is proven during Guth's lifetime, he is a sure bet to win a Nobel Prize.

In Guth's book (Guth, 1997), he reveals the unlikely path that brought him to inflation theory. In 1978, while a postdoc at Cornell, he attended a lecture by Robert Dicke on the topic of cosmology. Lectures on this were rare at that time, as cosmology was still coming into its own. The lecture was also attended by Guth's colleague Henry Tye, also a postdoc at Cornell. Tye was interested in a related problem in physics called the Grand Unified Theory (to unify, in a single mathematical model, the four known forces of nature), abbreviated as **GUT**. Tye convinced Guth to investigate connections between the Big Bang and GUT.

In the discussions about GUT that followed, Tye and Guth focused on the theory's prediction of a new particle called a magnetic monopole. These hypothetical particles, which have only a single pole (such as north) as compared to a bar magnet, which has two poles (both north and south), would provide a symmetry that was missing in the conventional theory of electromagnetism. The problem is that these particles were predicted to be so heavy (about a billion billion times the proton mass) that they could only have sprung into existence by the Big Bang. This problem appeared to have little connection with the concept of inflation theory. But sometimes the path to new theories takes a circuitous route. Such was the case here.

In thinking about the magnetic monopole problem, Guth was forced to also think about the questions surrounding the origins of the Big Bang. A visit to Cornell by famed physicist Steven Weinberg sparked the imagination of Guth, and he delved further into the study of connections between GUT and the Big Bang. Shortly thereafter, Guth went to a national laboratory near Stanford University that provided an ideal environment for more discussions about these rather esoteric ideas. Within a

---

[2] There are really several mathematical models for inflation, not just the one proposed by Guth. Here, I choose to focus on one of the simpler models.

few months, Guth had the basic mathematical framework for inflation theory. When his ideas were published in 1981, the main thrust of the paper is inflation theory and only mentions magnetic monopoles in subsection IV.

The point here is that theoretical physics is a collaborative endeavor, with lots of ideas coming from different people. In Guth's case, he was motivated by his colleagues to study GUT, and he focused on one aspect that didn't make sense: the production of magnetic monopoles in the Big Bang. By trying to solve that problem, it led him to a major breakthrough in cosmology that not only solved the monopole problem but also answered a number of questions about the origin of the Big Bang. Not every theorist will have such big impact, but centers of excellence staffed with people at all career stages seem to be a recipe for success.

To get started, what is a quantum fluctuation? The theory of quantum mechanics, which is now the basis of all modern physics, is well established. While this subject is not taught in beginning physics courses, quantum mechanics is essential to understand what happens at very small scales (such as the size of an atom or smaller). For example, the motion of electrons as they course through the transistors in your phone is dictated by the laws of **quantum mechanics**. Without an understanding of quantum mechanics, we wouldn't be able to design much of the machinery of modern technology, from lasers to computers.

However, quantum mechanics is very nonintuitive, meaning that it seems to go against much of our common sense. One of the laws of quantum physics is called the uncertainty principle (Heisenberg, 1927), which states that we can never make exact measurements of quantities such as energy and time. In particular, if we try to measure the energy of a system in a very short time period (say, the time it takes an electron to cross the nucleus), then there is an inherent uncertainty in that measurement, no matter how good our technology gets. It's a physical limit imposed by Nature. In the macroscopic world, we can get more precise measurements by creating better technology, such as using laser alignment tools rather than older tools like rulers and levels. Quantum theory tells us that there is a limit to how precisely we can measure *energy* during a short time period. If this seems strange to you, you are not alone. It defies common sense.

So, over a small period of time, the conservation of energy doesn't hold. If we measure all forms of energy over a longer period of time, then energy is conserved, but not over a very small interval like 0.00000000001 seconds. You might think, "Who cares?" Well, without this principle of physics, we wouldn't be here. The chemistry of life is governed by the laws of quantum mechanics. So, to think like a quantum physicist, you need to let go of some of your tenets about Nature. You need to believe in a subatomic world where particles can pop into existence and then disappear on timescales that are only relevant in the subatomic realm.

To answer the question we started with, a quantum fluctuation is when a particle pops into existence for a fleeting moment, then disappears just as suddenly. Quantum theory tells us that this happens all the time, even in empty space, for particles of very small mass.

The quantum fluctuation that sparked the Big Bang was, of course, very special, because that kind of fluctuation doesn't happen frequently. In fact, it's a once-in-a-lifetime event, where the lifetime is that of our universe. Quantum theory tells us that a quantum fluctuation of heavier particles is less likely than that of lighter particles. This suggests that the quantum fluctuation that sparked the universe would need an extremely heavy particle (if such a particle exists—see the discussion below) compared with the electron's mass. That fluctuation would be so exceedingly improbable that it almost never happens (say, once in about fourteen billion years).

Beware that there is a lot of speculation here. First, we are extrapolating the theory of quantum physics into a realm that hasn't been tested experimentally. Second, we don't know whether particles of such large mass actually exist. (Obviously, we can't produce such particles in our laboratory experiments. Otherwise, we might induce a second Big Bang.)

According to the inflation scenario, the type of particle that was created by a quantum fluctuation is from a classification of particles called scalar bosons and called the **inflaton**. This type of particle has very special properties and can be formed in an unstable state called a **false vacuum**. The false vacuum is a term unfamiliar to most people, and a simplified description of this is that the particle's mass and its potential energy are at a balance point that allows the particle to exist for a short while, then give off energy over time. It is somewhat like trying to balance a marble on top of a fixed bowling ball; the marble appears nearly stationary at first, but eventually rolls off, converting the potential energy from gravity to kinetic energy until it reaches the floor. Similarly, the quantum fluctuation creates a massive particle that rests in its own potential energy field, at a balance point where at first it slowly converts the potential energy to real energy (such as radiation), and then converts energy faster over time until it reaches the point of the "true vacuum" (the point where the potential energy bottoms out, see Figure 1.1).

The concept of a false vacuum is not just fantasy. Very similar mathematics predicted the existence of a particle called the **Higgs boson** (see Box 1.4), which was discovered in high-energy physics experiments (ATLAS, 2012).

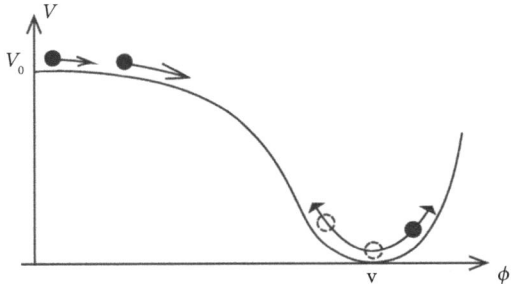

Figure 1.1 The concept of slow-roll inflation, where the potential energy (labeled V) has a plateau when the scalar field (labeled φ) starts off and rolls slowly down the curve until it reaches the "true vacuum."

Adapted from www.UniverseInProblems.com (2015). *Different models of inflation*, problem 14. See also Yu. L. Bolotin et al., arXiv:0904.0382, Chapter 8.

Named after physicist Peter Higgs, this particle (also of the scalar boson classification) was predicted by him (Higgs, 1964) almost fifty years before it was found in particle detectors at the huge multinational particle accelerator called the Large Hadron Collider (LHC). The mathematical theory used by Higgs, called quantum field theory, was originally developed to describe the physical mechanism responsible for radioactive decay, and the math naturally led Higgs to propose a hypothetical particle with a false vacuum that would eventually be verified by experiment. The Higgs boson differs from the kind of particle that sparked the inflation scenario, in that the Higgs boson has a small enough mass that it can be produced in terrestrial experiments, whereas the scalar boson of inflation is much heavier.

Box 1.4 The Higgs Boson

The Higgs boson is one of the fundamental particles that make up the Standard Model (SM) of particle physics (see Chapter 4). The importance of the Higgs boson is that it solidified theoretical ideas of how the electromagnetic force is connected to the weak nuclear force. The unification of these two forces, which don't appear alike in laboratory experiments, happens only at high temperatures (such as occurred in the Big Bang).

The idea of unification of forces is the holy grail of theoretical physics. Today, we know that the electric and magnetic forces are unified (by Maxwell's equations), but this was not always the case. Before electricity was widely available, the force

continued

Box 1.4 *continued*

---

between two magnets seemed very different from the force between two charged objects. The famed Maxwell equations showed that a moving electric charge (electricity) can generate a magnetic field and vice versa. Similarly, the weak nuclear force is responsible for radioactive decay, which seems very different from electromagnetism. Yet theoretical progress showed the two are connected.

The story of electroweak unification is fascinating. It exemplifies how theoretical physics is done. The foundations of the theory were laid by two physicists (Glashow and Salam) in 1959, but that theory had problems: it predicted several massless particles that were not observed in nature. A few years later, it was shown (by Higgs and Englert) that the theory could be modified in a way to avoid the massless particles, but then resulted in predicting a single massive particle (now called the Higgs boson). Another few years passed before Steven Weinberg demonstrated that this theoretical framework could be used to unify the weak and electromagnetic forces, giving specific predictions for experimental measurements. Theoretical physics is collaborative, with each group building on the work that came before. Several Nobel Prizes were awarded for these achievements.

Weinberg's predictions were dramatically verified at a new high-energy accelerator facility (located in Geneva, Switzerland) in 1983. The discovery of the Higgs boson would wait for another thirty years, until 2012. It took that long for improvements in technology (and to get the worldwide financial support to build higher-energy accelerators). But once the Higgs was seen experimentally, it verified the predictions formulated by Higgs and Englert. The ideas presented there were also used in other theories, such as the Grand Unified Theory that Alan Guth and others used when formulating inflation theory (see Box 1.3).

Today, the theory of electroweak unification is firmly established. It gives some hope that the next step can be reached: to unify the electroweak forces and the strong nuclear force. (The latter holds together protons and neutrons in the atomic nucleus.) If that's accomplished, then we can predict how various subatomic particles interacted at the earliest moments of the Big Bang. That, in turn, would help us understand how the universe came to be. Only time will tell whether such theoretical progress can be achieved.

In a nutshell, quantum mechanics allows particles to fluctuate into a brief existence, and the particle that sparked the Big Bang is conjectured to be a massive particle with special properties (a false vacuum) that make its probability of fluctuation a once-in-a-universe event. I emphasize that a detailed understanding of why this fluctuation is so rare is not possible with our present knowledge. An explanation should come with further experimental and theoretical work.

The reason that the particle must be created with a false vacuum is that this scenario allows the mathematics of general relativity to create a very bizarre world, where the gravitational field pushes rather than pulls. Our common sense tells us that gravity always pulls on an object. A physicist might say that gravity is "an attractive force," meaning that two masses are attracted (pulled together) by the force of gravity. However, general relativity allows space-time (which, recall, is the same as gravity) to curve in such a way as to make a repulsive force, meaning that two masses are pushed apart under this unusual situation. This is sometimes referred to as **negative gravity** (Guth, 1981). This situation is exactly what happens in a tiny region of space filled with an inflaton particle. Then space itself is pushed apart, or inflated, as the particle transitions from its original state (with a false vacuum) to that of the true vacuum.

As an interesting side note, Einstein also noticed that his equations allowed gravity to push as well as pull. Einstein believed in a static universe, one that is not expanding or contracting. Einstein's theory was published before there was evidence that the universe is expanding, and at that time most people believed in a static universe. But if gravity can only pull, this predicts that (from a static starting point) that the universe would eventually collapse in upon itself in a Big Crunch. Einstein realized that his equations allowed there to be an additional constant (Einstein, 1917), now called the **cosmological constant**, that would balance the pull of normal gravity with a push of negative gravity. In this way, he could mathematically describe a static universe. A decade later, when Edwin Hubble published his observations that the universe is expanding, Einstein regretted his publication introducing the cosmological constant. Some say he called this the biggest blunder of his career (although it's not clear that he ever said these exact words). The main point is that Einstein's equations allow for a term that expands space.

The bizarre world of inflation is difficult to describe using just words. Exactly how this happens is, of course, tied up in the math of general relativity. The key thing to understand here is that the laws of physics allow such a thing to happen, even if the probability of its happening is once in fourteen billion years or more. There are other things in daily life that are allowed by the laws of physics but don't happen because the probability is so low. For example, it's possible to flip a coin randomly and have it come up heads a hundred times in a row. The probability for this is not zero, but it's so small that it would likely not happen in the age of the universe.

1.2 Abracadabra: Something Comes from Nothing

Getting back to the Big Bang, the hypothesis is that it had a spark (the quantum fluctuation) followed by a short period of inflation (due to the negative gravity) where space pushes itself apart. Now a miracle appears to happen. Recall that it takes energy to curve space-time. What happens when space-time un-curves, as it does when it inflates? (Think of a balloon, which has a sharply curved surface when small, but when blown bigger the surface becomes flatter, or has less curvature, as shown in Figure 1.2.) The answer is that space-time gives off energy as it inflates. The longer the inflation, the more energy it releases. This energy can turn into matter (recall $E=mc^2$), and matter is what we have in our universe.

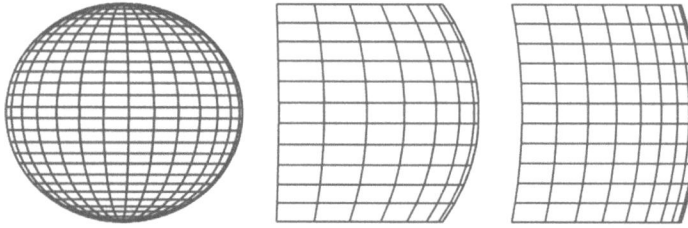

Figure 1.2 View of a sphere from afar (left), closer (middle), and near the surface (right). The surface appears flatter and flatter as one gets progressively closer to the surface. Adapted from Guth (1997).

The big question is: can enough energy be produced to account for all matter in our universe? Here is what Alan Guth said when interviewed about this (Edge, 2002):

> Inflation is an attempt to answer that question, to say what "banged," and what drove the universe into this period of enormous expansion. Inflation does that very wonderfully. It explains not only what caused the universe to expand, but also the origin of essentially all the matter in the universe at the same time. I qualify that with the word "essentially" because in a typical version of the theory inflation needs about a gram's worth of matter to start. So, inflation is not quite a theory of the ultimate beginning, but it is a theory of evolution that explains essentially everything that we see around us, starting from almost nothing.

I want to emphasize one point from the above quote. Starting with a quantum fluctuation that has about 1 gram of mass (that's about a paperclip), then "essentially all the matter in the universe" results from inflation. This sounds preposterous! How can it possibly be true? This goes against all common sense. Yet the equations of physics show that this assertion is correct, assuming general relativity can be applied at very small distances (along with a few other assumptions, such as the existence of a very massive scalar boson or some alternate way to create a false vacuum).

The math behind the inflation scenario has been verified by some of the best minds on the planet. The trick here is that general relativity allows a tradeoff between curvature of space-time and energy. To produce a whopping amount of energy, you need a corresponding gigantic change to the curvature of space-time. Guth has remarked that inflation is "the ultimate free lunch" (Guth, 1997). Let's explore how this happens in more detail.

First, I want to make a general statement of *caveat emptor* (buyer beware). To really believe that the inflation scenario is correct, we need a theory of quantum gravity, which currently doesn't exist. We have a theory of gravity (Einstein's equations of general relativity), and we also have a theory of quantum mechanics, but we don't know how to merge the two theories together. Until someone comes up with a theory of quantum gravity, the inflation scenario is just a hypothesis where Einstein's equations are extrapolated down into the quantum realm, without proof that they can be applied there. However, if we allow this assumption, then we'll see that the inflation scenario makes predictions that agree extremely well with multiple astronomy observations.

In the inflation scenario, the early universe started in a tiny fraction of a second, in an incredibly small volume. In Figure 1.3, the radius of the universe is shown versus time after the start (after the quantum fluctuation). The numbers shown are in standard "factors of ten" notation, where, for example, $10^3 = 1,000$ and $10^{-3} = 0.001$. Inflation starts at about 10^{-35} seconds to about 10^{-33} seconds, during which the radius increased in size by a factor of about 10^{40}, going from, say, a size of 10^{-45} meters to 10^{-5} meters (the exact size at the start inflation is not important as long as the radius increases by many orders of magnitude). In other words, the inflation scenario has the universe expanding in a tiny fraction of a second, increasing in radius a factor of ten many times over. The thought that this could really happen is mind-boggling. But the gigantic change to the radius means an equally big change to the curvature of space-time, resulting in (according to Einstein's equations) the release of a whopping amount of energy. This energy creates the mass of the entire universe.

To spark inflation, we need a quantum fluctuation that has a mass of about 1 gram. According to a formula from quantum mechanics (called the uncertainty principle), the time that such a fluctuation would last is about 10^{-33} seconds, long enough to start inflation. However, the probability for such a

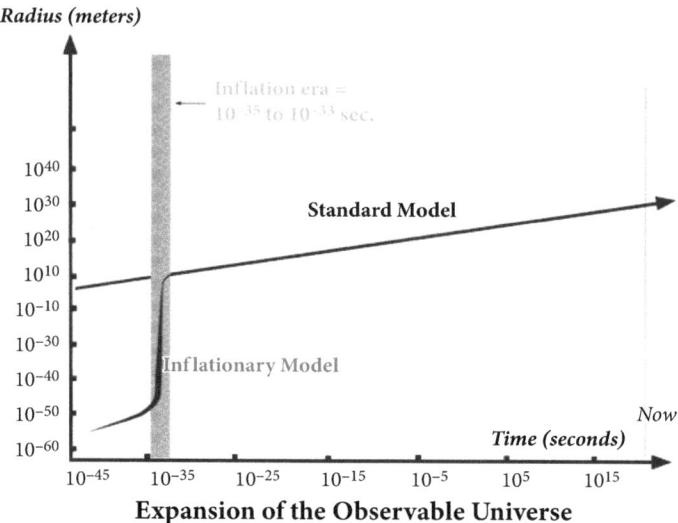

Expansion of the Observable Universe

Figure 1.3 The size of the universe in powers of ten (units of meters) versus the time from inception in powers of ten (units of seconds) predicted from the inflation scenario. The model of the Big Bang is a straight line, whereas the inflationary model suddenly jumps in size by many orders of magnitude in a very short time.
Reproduced from Wikimedia Commons (author unknown, public domain, 2016).

fluctuation to occur in empty space is so small that it almost never happens. Lucky for us that it happened once! Could it happen again? Could a second Big Bang wipe out all of civilization? Perhaps, but the fact that we're still here suggests that it's a very unlikely event.

What makes inflation stop? That's a more difficult question to answer, because here we need to extrapolate into regions of energy density that we can't yet measure in experiments. The current thinking is that the infant universe undergoes a "phase change." The idea of a phase change is something that happens to ordinary matter, such as water changing into ice. Water is the fluid phase, and ice is the solid phase of the same molecules, H_2O. When water turns to ice or vice versa, we call that a phase change. Experimentally, when a phase change occurs, heat is either given off or absorbed, depending on which direction it goes (water to ice or ice to water). In the case of inflation's phase change, something about the state of matter created by inflation needs to change as the universe cools. That phase change is hypothesized to happen when inflation's scalar boson drops into its true vacuum state (the lowest point of potential energy in Figure 1.1).

Some as-yet unmeasurable phase transition likely happened in the early universe, providing excess heat that stops inflation. This is still speculative, but the fact is that something had to change to stop inflation. Even though the exact physical mechanism of inflation's phase change is unknown, due to our inability to build experiments at the incredibly high energies present in the early universe, the concept of a phase change is not unfamiliar (see Box 1.5). In the inflation scenario, we take this concept and extrapolate it. Nature has a habit of repeating itself, so it seems reasonable to assume that this is what stopped inflation.

Box 1.5 The Quark-Gluon Plasma

A phase change occurs when nuclei are smashed together at high speed. The transition is analogous to what happens when ice is heated. In ice, molecules are confined in place as ice crystals. When heated, though, they are suddenly free to slosh around as water. Another example is when the electrons bound to atoms are heated (by electrical discharge) and the electrons can move around freely as an ionized plasma. Similarly, tiny particles called quarks are confined inside protons until heated (by high-energy collisions) and then become free to move around. This deconfinement of quarks is called a quark-gluon plasma (QGP).

Quarks (see Chapter 4) are fundamental particles in the SM of particle physics. In contrast, neither protons nor neutrons are fundamental particles, as both are made up from combinations of quarks. The quarks are tightly bound together, so the proton appears as a self-contained particle. When struck by high-energy electrons, however, the quarks inside the proton become evident. Gluons are also fundamental particles but are different from quarks in several ways. Gluons, the carriers of the strong nuclear force, are exchanged between quarks much like a baton is exchanged between two runners. The equations that describe this process are well established.

When quarks are heated, becoming free, the gluons swarm around the quarks like a fluid. The phase transition from an ordinary nucleus (with protons and neutrons) to the QGP has been studied using particle accelerators. These experiments help us to understand how subatomic particles interacted in the first moments of the Big Bang. In a sense, we can create a "little bang" in the laboratory, although this only re-creates the conditions in the moments after inflation. These experiments are carried out at the Relativistic Heavy Ion Collider (RHIC) in New York and at the Large Hadron Collider (LHC) in Switzerland.

continued

Box 1.5 *continued*

The discovery of the QGP phase transition is somewhat recent (in the past decade or so), and physicists are just now starting to learn about the properties of this new state of matter. Before these experiments, using large-scale particle detectors, only theoretical guidance was available for what occurred during this part of the Big Bang. The new experimental work has provided several surprises about the QGP that were not predicted by existing theories. For example, the QGP flows like the most perfect liquid, with nearly zero viscosity. Yet quarks, as they pass through the QGP, interact strongly with the QGP (transferring their kinetic energy and thus heating the QGP).

The bottom line is that the properties of this new state of matter are surprising, showing that much can be learned from phase transitions. Of course, the phase transition that ostensibly occurred during inflation is still a mystery, but advances in both theory and experiment may yet shed light on the early stages of the Big Bang.

1.3 Inflation at the Tipping Point

Discoveries in particle physics are now placing limits on the physical properties (such as mass) and interactions among heavy particles (such as how the Higgs boson couples to the heaviest known quark). This increases our knowledge of how quantum fluctuations of a scalar boson (such as the Higgs) occur, and calculations can be compared with experimental results. Using this mathematical framework, we then boldly extrapolate to higher mass particles. In an article in the journal *Physics Today* (Lykken and Spiropulu, 2013), the authors say:

> The possibility that the universe could be in a metastable vacuum has been studied since the 1970s, but only now can scientists plug in the numbers. Taken at face value, the result implies that eventually (in 10^{100} years or so) an unlucky quantum fluctuation will produce a bubble of a different vacuum that will expand at nearly the speed of light, destroying everything. Strikingly, the measured Higgs and top-quark masses put the universe right at the edge of the stability versus metastability divide; if the Higgs boson were a few percent heavier, or the top quark a few percent lighter, then the vacuum would be stable. Is our existence at the edge just a coincidence, or is Nature telling us something?

Nature seems to be at a balancing point, where a quantum fluctuation capable of starting a new universe could happen, but only very rarely. If the universe were stable, then such a quantum fluctuation could not happen; if it were less stable, it would happen too often. Why should the masses of the Higgs boson and the top quark be so perfectly balanced (see Figure 1.4)? I'm sure no one can answer that question today, but perhaps a deeper theory (such as the untested theoretical idea of particles being "strings" in a higher-dimensional space, called string theory) will eventually lead us to understand why our universe exists at a balance point (i.e., a metastable state). For now, it's one more example of how our universe is slightly off-balance, allowing it to exist long enough for human life to develop.

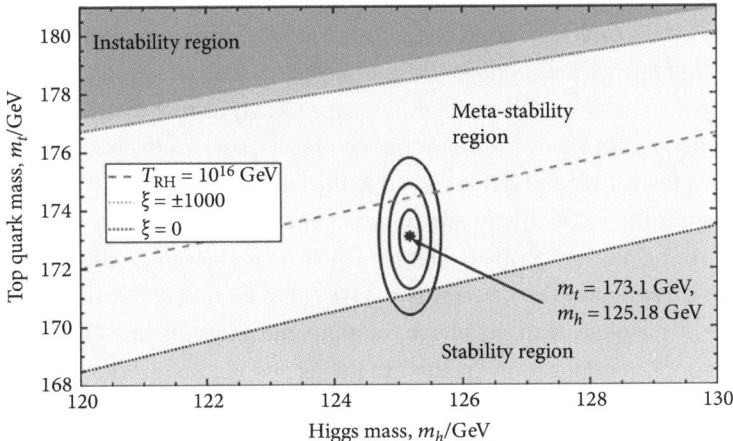

Figure 1.4 Plot of the mass of the top quark versus the mass of the Higgs boson. The point in the center of the bullseye is at the measured values, which lies in the region of a metastable universe.

Reproduced from Makkanen (2018). Via Wikimedia Commons (CC BY 4.0).

1.4 Evidence for the Inflation Scenario

Up until now, the discussion has been rather hypothetical, without observational evidence (except the fact that our universe exists). Of course, Einstein's equations have been carefully tested by experiments, as has quantum mechanics, but can the inflation scenario also make predictions that can be tested? The answer is yes, in the sense that we can observe aspects of the universe (such as the "afterglow" of the Big Bang) that are consistent with this scenario. Obviously, we can't do experiments that re-create the conditions during inflation, but it's certainly possible to identify what present-day observations can be explained by inflation (see Box 1.6).

Box 1.6 Inflation Simplified

Let's look at a much simpler scenario than inflation. Say I throw a ball up into the air. We all know what happens: the ball goes up to the top of its arc, stops momentarily, and then starts back down. Clearly, the ball started with some energy (kinetic energy). When the ball stops at the top, though, what happened to that energy? A physicist would say that the energy has become "gravitational potential energy," which means that the energy went into the ethereal "gravitational field," which we can't see, hear, or touch. Tossing a ball up is so familiar that we never really stop to think about how bizarre it is that energy seems to disappear. We "know" that there's a tradeoff of energy between kinetic energy (which we can see) and energy of the gravity field (which we can't see). Common sense tells us that there must be a gravitational field, since we can see its effect on objects, such as the ball. Energy comes out of the gravity field and is transferred to the ball as it drops.

In a sense, the inflation scenario should not seem so bizarre either if you know Einstein's principle that gravity is related to the curvature of space-time. For inflation, which un-curves space-time, energy is released (and some of that energy is converted to matter). In analogy with the above example, the gravity field (aka curved space-time) is supplying the energy needed to power the Big Bang. What is different (in fact, very different) from the above example is the mechanism by which the energy is released. In the inflation scenario, this energy comes via a subatomic particle (a scalar boson) that has its own potential energy "field."

In other words, we need a scenario where gravity can push rather than pull, and mathematically this is possible if the scalar boson is born with a potential energy curve having a false vacuum (similar to what theoretically happens when a Higgs boson is produced by high-energy collisions of two head-on protons). Even though the mechanism is very different, the idea that a changing gravity field can supply energy should not be such a foreign idea. Is it so strange that a ball gets energy from the gravitational field when it starts to drop?

Once you are comfortable with the concept that space-time can be stretched and compressed, then the idea that so-called dark energy can be present in the universe seems not so strange. The use of "dark" here is not a good adjective, and it might be better called negative energy because it pushes space-time apart (the opposite of gravitational attraction). You may have heard that dark energy is present in our universe, and this is deduced by comparing theoretical models of an expanding universe to astronomical observations showing that far-off galaxies are accelerating away from each other (i.e., not attracting). In fact, these theoretical models fit a variety of observations, including the CMB and clustering in the positions of galaxies (called baryon acoustic oscillations, or BAOs). Unless something is very wrong with the models, dark energy is enmeshed in the structure of space-time for our slightly off-balance universe.

One of the problems that the Big Bang (without inflation) could not explain is called the horizon problem. In this context, horizon means looking at to the edge of the universe. We want to view light from as far away as we can see. Here, I generalize the use of "light" to mean any wavelength of electromagnetic radiation, from long-wavelength radio waves to short-wavelength X-rays.

Visible light, as most people know, is just a small slice of the full spectrum of electromagnetic waves. We can feel infrared light but can't see it. However, we can build devices (infrared cameras) that can see infrared light. Similarly, we can build telescopes that see the remnant heat from the Big Bang, which occurs at wavelengths of microwaves. This light, known as the **cosmic microwave background** (CMB), was discovered accidentally (Penzias and Wilson, 1965). At the time, those scientists thought the CMB was just noise in their microwave receiver, and they tried to get rid of this "background" to enhance the signal they were trying to measure (see Box 1.7). The remarkable thing about the CMB is that it's there in the sky all the time and in any direction. They didn't know about the Big Bang or the fact that there would be remnant radiation from the Big Bang (the CMB) that we can see today.

Box 1.7 Penzias and Wilson

Arno Penzias and Robert Wilson worked at Bell Labs on supersensitive microwave receivers, for both communications use and astronomy observations (of known radio sources in the sky). Their "telescope" was about 50 feet long and in the shape of a horn (to amplify the microwaves). When testing it, they noticed a "hiss" of microwave noise coming from the sky. The noise was spread evenly across the sky and about a hundred times stronger than they expected. Little did they know that this "noise" was the CMB and would win them a Nobel Prize!

In trying to diagnose the cause of the noise, they clearly ruled out any terrestrial source as well as other celestial sources such as the sun and the Milky Way (since the horn could be rotated around its axis). At one point, they went into the horn and noticed that there was a white substance, later identified as pigeon droppings; cleaning that out didn't change the noise. At that point, they realized the only reasonable

continued

Box 1.7 *continued*

explanation was that the noise was really a signal coming from across the sky. This made no sense to them because, at the time, there was no physical mechanism that would generate microwaves uniformly from everywhere out in space.

If your eyes could detect microwaves, you would see a background glow to the night sky at a particular "color" corresponding to a wavelength of 7.35 cm. It would be the same regardless of where you looked in the sky, both day and night. What could be causing this? If it were coming from stars, you would see point sources, much like you see with visible light. The microwaves aren't just coming from point sources. To find out what it could be, Penzias and Wilson started talking with astronomers.

A friend of Penzias, a professor at MIT, told him about a preprint (a paper that is circulated privately before publication in a peer-reviewed journal) from astronomers at Princeton University who had calculated that there should be a background "glow" as a remnant from the Big Bang. Since an explosion expands outward in all directions, some of the glow from one part in the expansion should radiate out to other parts. The location of Earth is not special and started out somewhere inside the Big Bang. So, a prediction of the Big Bang is that microwaves should be visible in all directions.

Penzias and Wilson now had an explanation for the noise. Their measurements were not known to the Princeton astronomers, and similarly the idea of the microwave background was not known to Penzias and Wilson when they first took data. It was a perfect merging of experimental data with theoretical predictions. The Big Bang predictions matched very nicely with the data of Penzias and Wilson. But it took several decades of work (with space satellites built specifically for this purpose) to get more precise data that showed just how incredibly good the predictions are from the Big Bang theory.

The story of how the CMB came to be understood as a remnant of the Big Bang has been told many times, and I won't repeat it here. But one point that is critical to understand is that the distribution of microwaves, their intensity at each wavelength, matches exactly with what we should expect to see based on other heat sources.

For example, the sun radiates light over a range of wavelengths, which we see as colors. The most intense wavelength that reaches our eye corresponds to yellow light, but we can also feel (as heat) the less-intense wavelengths of infrared light and see the effects of ultraviolet light (giving sunburn). If you were to measure the intensity of light from the sun at each wavelength, you would find a characteristic shape called a blackbody spectrum. The name is just historical (the sun is not black), but the point is that any object that heats up will give off a distribution of wavelengths with intensities that can be calculated mathematically, having the blackbody shape. The CMB has precisely this shape. For this reason (and others), the CMB is associated with the remnant heat from the Big Bang.

Now comes the peculiar aspect of the CMB, which gives us the **horizon problem**. If you look at the sky in one direction (e.g., east), the spectrum of the CMB is exactly the same as if you look in the opposite direction (e.g., west). At first, you might be tempted to say, "Of course, it should be that way," because the Big Bang went out in all directions. However, there's a more subtle problem here. The CMB travels to you at the speed of light (as do all electromagnetic waves), and the CMB coming from the east is from the horizon, going straight into your detector. Since nothing can travel faster than the speed of light, the east horizon can't know about the spectrum at the west horizon. Yet both have an identical spectrum. How can one horizon know about the other horizon?

To give an example of how strange this is, imagine being in a large grass field with two friends, and none of you have a cell phone (or other communication device). The only way to communicate is by sight or sound. Your friends stand on opposite sides of the field, too far away to see small details of each other, but you have some binoculars. At your signal, each friend randomly chooses a card from a full deck and holds it up. You look at one friend with your binoculars, then the other, and both are holding up exactly the same card, say a three of clubs. How could this be if they hadn't communicated with each other? The answer is obvious: they must have met earlier and exchanged information, a plan to hold up the same card.

In the same way, the light of the CMB from one side of the universe (from looking east) can't be exactly the same as that from the other side of the universe (looking west) unless the two sides had been in contact at a previous time. In the standard Big Bang scenario, it was just assumed that everything started from an infinitesimal point and the expansion is uniform. But why should the expansion be perfectly uniform? Quantum mechanics tells us that there are random fluctuations, and this implies that the expansion has nonuniform regions. So, particles traveling in one direction out from the Big Bang don't need to be exactly the same velocity (or temperature) as particles traveling in the opposite direction, due to quantum randomness. Should we throw away quantum physics in order to explain the CMB?

The problem of uniformity is solved by the inflation scenario (Guth, 1981). During the expansion phase, negative gravity pushes space apart faster than the speed of light. At first, this might sound like it violates the principle of special relativity, that nothing can go faster than the speed of light.

However, the correct statement is that nothing travels through space faster than the speed of light. Special relativity assumes that space doesn't expand. That's where general relativity enters. Under the equations of general relativity, space can expand, and in fact, it's possible for space to expand faster than light. What this means is that two parts of space that were in contact before inflation (before 10^{-36} seconds) are now separated after inflation ends (after about 10^{-33} seconds). This doesn't sound like much; when you work out the math, though, it actually explains the CMB spectrum being the same from opposite directions. In other words, inflation solves the horizon problem that was an inconsistency of the Big Bang theory.

A second problem of the Big Bang is the so-called **flatness problem**. By this, we refer to the curvature of space-time today, and what this implied about the curvature of space-time at the start of the Big Bang. The curvature of space-time is best explained by analogy, in terms of a ball shot by a cannon on Earth. The example I'll give comes from Isaac Newton, who first described mathematically the force of gravity.

Think of shooting a cannonball from a mountaintop. What trajectory does it take? For simplicity, let's ignore the air friction that acts to slow the cannonball. Gravity curves the trajectory of the ball. If the ball has a slower speed, then it falls to Earth closer. If it has a medium speed, it goes farther, but still eventually falls to Earth, as shown in Figure 1.5.

But if it has just the right speed, it will fall at the same rate that the surface of the Earth curves, going around Earth in an orbit. What happens if we fire the cannonball even faster? With enough speed, it will escape the Earth's gravity, flying off into space. Only if the ball has exactly the right speed can it orbit in a perfect circle (see Figure 1.5).

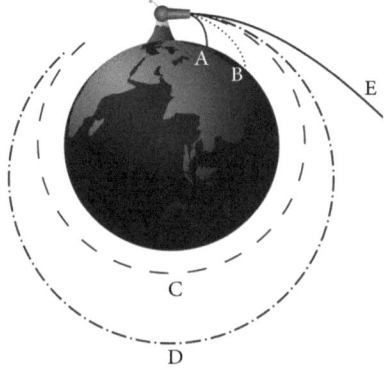

Figure 1.5 Schematic diagram of Newton's idea. If a cannon could be put on a mountain high enough (above the atmosphere) and could fire its cannonball fast enough, it could orbit the Earth, as shown by curves C and D.

Reproduced from B. Brondel (2010). Via Wikipedia Commons (GNU Free Document License).

Let's now change the picture slightly and fire the cannonball straight up. If it doesn't have enough speed, it will eventually fall back to Earth. If it has too much speed, it will go forever away from the Earth, keeping a significant speed. But between these two extremes is that case where the ball continues moving away, losing energy due to gravity, with its speed coming closer and closer to zero but never quite stopping. You would need to be extremely lucky to get the speed just right, to exactly balance the initial (kinetic) energy of the cannonball against the gravitational (potential) energy of the Earth. (For this example, I'm making the simplifying assumption that the ball is not affected by any forces other than the Earth's gravity.)

Could you balance it sufficiently well so that it goes out for ten years before stopping? How about going for a million years before stopping? Or even better, could you get the speed just right to go fourteen billion years (close to the age of our universe) before stopping?

This is similar to what we need to assume for the Big Bang for the flatness problem. We observe that all galaxies are moving away from each other, as you'd expect for the objects in an explosion. The speed of the galaxies (along with other information) shows that our universe is in a nearly perfect balance, where all sources of energy (kinetic energy, gravitational energy, and so on, including the much-reported but not-yet-understood "dark energy") combine to exactly cancel, just like the case of the cannonball above. The point is that our universe has been expanding outward for almost fourteen billion years and slowing down for most of that time. It's as if the Big Bang had a magical ability to get the force of the explosion just right so that the galaxies don't fall back together into a Big Crunch, but also slow down, never reaching zero.

Box 1.8 Energy Density

Since Einstein tells us that energy and mass are interchangeable ($E=mc^2$), let's talk about a more familiar quantity: mass density. For example, lead is denser than wood, and wood is denser than Styrofoam. The density of a material is measured in kilograms (kg) per cubic volume. The paper of this book has a mass density of about 1,000 kg per cubic meter. A cubic meter is about the amount of paper that fits on a palette in a warehouse, and a forklift is needed to lift it.

To calculate the mass density of our galaxy, we use the same method. So, we need a good estimate for the mass and the size of the Milky Way galaxy. There's a huge amount of empty space between stars. So, even though a star (including the planets orbiting it) has a large mass, the mass *density* of our galaxy is rather low.

continued

Box 1.8 *continued*

Astronomers say that our galaxy contains billions of stars. From astronomical observations, a good estimate of the **total mass of stars in our galaxy is about 10^{41} kg.** That seems like a lot. Remember, though, that density is defined as mass divided by volume, and the galaxy's volume is very big, too.

You might be asking yourself "How big is the galaxy?" This is a harder question because there is no clear edge to it. However, reasonable estimates give the diameter of the Milky Way disk as about 100,000 light-years, where a light-year is the distance that light travels in one year. Converting this to meters, one light-year is approximately 10^{16} m.

As a next step, consider the surface area of our galaxy's disk. Using the equation for the area of a circle (πr^2, where r is the radius of the disk), **the surface area of our galaxy is about 10^{42} m².**

To compare this with paper, let's do a "thought experiment." Imagine that you could spread paper, one sheet thick, across the galaxy. Paper is about 0.1 mm thick. Using the mass density of paper (given above), 1 square meter of paper has a mass of 0.1 kg. So, 10^{42} square meters (m²) of paper would have a mass of 10^{41} kg. Comparing this with the total mass of our galaxy, it's the same.

What this means is that if we could spread evenly all of the star mass of our galaxy into a thin layer (the "peanut butter" approach) of 0.1 mm thickness, then it would have the same density as paper!

Combining this with the fact that galaxies are separated by distance of billions of light-years, you can see that the mass density of normal matter in our universe *at present* is extremely low compared to the mass density of materials on Earth. Of course, when the universe was young, and the size of the universe was smaller, the mass density was much higher.

For those who like to learn by doing calculations themselves, here are a few facts:

Total mass of the Milky Way galaxy = About 50 billion solar masses
1 solar mass = About 2×10^{30} kilograms (kg)
Speed of light = About 3×10^8 meters/second = 300,000 kilometers (km)
1 year = About 3.1×10^7 seconds
1 light-year = About 1 trillion km
1 km = 1,000 meters (m)
0.1 millimeters (mm) = 10^{-4} m
Mass = (Density) × (Volume)

For simplicity, this calculation ignored dark matter and dark energy, which also contribute to the energy density of our universe. For the above example, only familiar matter made of atoms (called baryonic matter) was used. Including inter-stellar gas and dust in the galaxy's disk, this changes the calculation by just 10–15%.

The full picture, to account for all sources of energy, is more complicated than just accounting for kinetic and gravitational energy. In fact, it takes a very long time (and many words) to explain how professional astronomers and cosmologists have come to this conclusion, so let's just summarize the results. Cosmologists use the Greek letter omega (Ω) to express the **critical energy density** of the universe (see Box 1.8). If $\Omega=1$, this corresponds to a perfect balance point. If Ω is slightly greater than unity, then our universe would already have ended in a Big Crunch. However, if Ω were slightly less than unity, the energy density would be less (matter becomes more diffuse), and this leads to less gravitational pull and so galaxies fly apart (see Figure 1.6). Extrapolating back in time, the math shows that the Big Bang had to have $\Omega=1$ to a precision of one part in a million billion! In other words, $\Omega=1.000\ldots$ with at least fifteen zeros following the decimal place (Guth, 1997). How could such precision be possible? Since $\Omega=1$ is for flat space, that's why it's called the flatness problem. Is space-time perfectly flat or slightly off-balance?

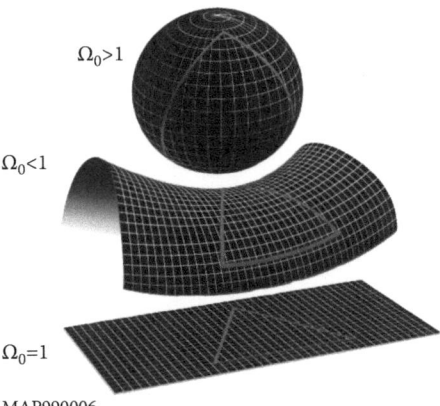

$\Omega_0>1$

$\Omega_0<1$

$\Omega_0=1$

MAP990006

Figure 1.6 Shapes of space-time for positive (top), negative (middle), and flat (bottom) curvature, corresponding to critical energy density values of $\Omega > 1, \Omega < 1$ and $\Omega = 1$, respectively. The triangles show that geometry is different on each surface.
Reproduced from NASA/WMAP Science Team (2006). Via Wikimedia Commons (public domain).

The inflation scenario solves the flatness problem (Guth, 1981). In the inflation scenario, space inflates uniformly, driving space-time to be flat. Think again about the analogy with the inflating balloon given in the previous section. Any small wrinkle on the surface gets flattened out as the balloon inflates. A similar process happens to Ω, the parameter that corresponds to the flatness. Even if it was not exactly unity before inflation, due to quantum effects, then after inflation the math shows that Ω gets very close to unity. If there is another way to get this to happen (other than inflation), to get a universe with exactly the right critical energy density so that it lasts for fourteen billion years, then it would be a very hard sell. The most natural explanation is the inflation scenario.

Even though a solution to the horizon problem and the flatness problem is a good reason to adopt the inflation scenario as real, other astronomical observations are consistent with this modification to the Big Bang theory. For example, one can calculate the quantum fluctuations expected before inflation (Brandenberger, 1985) and look for these effects in the universe after inflation. We might expect very slight differences (about one part in 10,000) to the CMB spectrum coming from different parts of the sky. This has been measured with high accuracy by the **Planck satellite**, a mission of the European Space Agency (Aghanim et al., 2020). A map of the sky showing the CMB temperature is shown in Figure 1.7 (after removing non-CMB effects, such as the background from the Milky Way). You can think of this map as a wide-angle camera view, with 360° from left to right and north pointing up. The colors represent the small fluctuations, magnified by a factor of about 10,000. For example, the darker (blue) regions are cooler by about one part in 10,000, compared with the global average, and the lighter (reddish) regions hotter by the same amount. I emphasize that this is the temperature of the sky, the remnant heat of the Big Bang. The size of these fluctuations supports the inflation scenario (Linde, 1990).

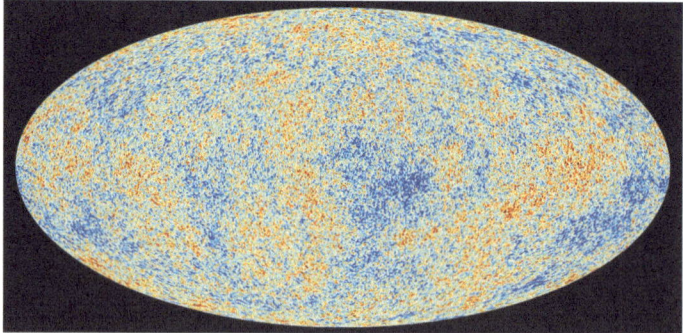

Figure 1.7 Map of the cosmic microwave background (CMB) in a 360° view of the sky, where the equator corresponds to the plane of the Milky Way. The different colors represent tiny variations of the CMB temperature seen from all directions in the sky. Reproduced from ESA and the Planck Collaboration (2023). Via Wikimedia Commons (CC BY 4.0).

Even more impressive is the quantitative information that can be extracted from this picture. You can see that some darker regions are bigger than others. One can use a computer to measure the size of each dark region and, similarly, the size of each light region. By counting the frequency of each size, one gets a so-called power spectrum, which is shown below. You can see that the average size of a region of fluctuation is about 1°, which matches what your eye sees from the above CMB temperature map. What your eye probably can't see on the map, but becomes clear from the computer's counting, is that there are additional peaks in the power spectrum, at smaller angular scales (see Figure 1.8).

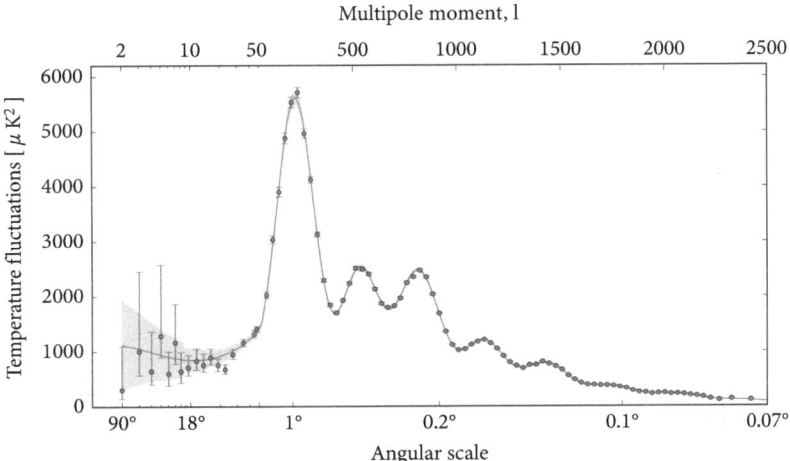

Figure 1.8 The so-called power spectrum of the CMB as measured by the European Space Agency (ESA) and the Planck Collaboration. The horizontal scale is the angular separation in the sky of the temperature fluctuations (see Figure 1.7), showing a peak when the fluctuations are about 1 degree. This spectrum comes from fits of multipole moments (top scale) to Figure 1.6.
Reproduced from ESA Multimedia Images (2013). Via www.esa.int (ESA Standard License).

The first peak in the CMB power spectrum, at about 1°, tells us that we live in a universe that has space-time curvature that is flat (Plionis, 2002) with $\Omega=1$. The ratio of the first and second peaks tells us that "ordinary" matter, elements from the periodic table having protons and neutron in the nucleus, accounts for only about 4% of the mass necessary to make the universe flat (Akrami et al., 2020). The third peak tells us that **dark matter** exists, meaning there is a particle that's different from (and doesn't interact with) ordinary matter and permeates the universe. Dark matter accounts for about 23% of the energy density.

To get to $\Omega=1$, we need more energy density, since ordinary matter and dark matter together only account for about 27% of it. To fit with the evidence, which includes measurements of the brightness of distant supernova in far-off galaxies, the mathematical models using Einstein's equations require additional energy density in the universe, which we call **dark energy**. One interpretation of dark energy is that it's the energy of empty space, which confounds the meaning of what we mean by "empty." Whatever it is, even if it goes against common sense, we need to accept that it's there if the observations are correct (and the mathematical models based on general relativity are correct).

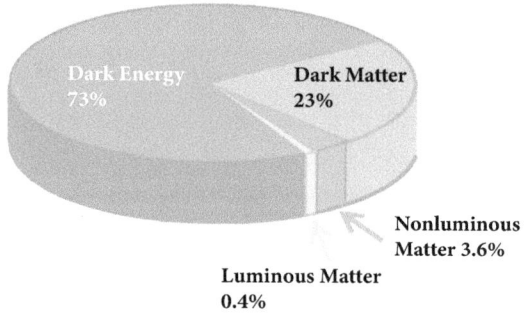

Figure 1.9 Pie chart of the universe's mass coming from normal matter (both luminous and nonluminous), dark matter, and dark energy, based on comparison of models of an expanding universe to a variety of measurements.
Reproduced from B. O'Hare (2008). Via Wikimedia Commons (CC BY 3.0).

The current picture, with dark energy making up more than 70% of the universe's energy density (see Figure 1.9), leads to the scientific models (based on general relativity) to predict that the universe is expanding. This is in accord with evidence of telescope observations of supernova (type Ia) seen in very distant galaxies. This has interesting consequences, such as far-off galaxies disappearing from view over long timescales (Krauss and Scherrer, 2007).

Of course, one could invent a new set of equations to replace Einstein's general relativity and the inflation scenario. Some cosmologists have done this, such as assuming the force of gravity differs over long distances, but none of these alternative scenarios have withstood the test of time. In contrast, Einstein's equations continue to be tested with a variety of precision experiments, and so far every prediction (calculated in advance of the measurements) of Einstein's equations has turned out to be correct (Will, 2014). Furthermore, the inflation scenario has made predictions that can be tested with future observations, and astronomers are working hard to either verify or discredit such predictions. Until a better theory emerges (or until the inflation scenario is discredited), the current interpretation of a universe filled with ordinary matter, dark matter, and dark energy will stand.

1.5 Summary

The Big Bang theory had some problems, because measurements showed that the universe was flatter and more uniform than anyone expected. In addition, where did all the mass of the universe come from? Furthermore, quantum theory predicted that there should be density fluctuations in the early universe that could be observable today as a slightly nonuniform CMB temperature. An exquisite precision was needed to balance the energy density at $\Omega=1$. All these problems are solved in the inflation scenario of today. It remains to further test the predictions of inflation theory, which is in progress.

Sometimes we need to set aside our common sense when we're dealing with Nature in a regime where our everyday experience doesn't apply. Whether that's at the subatomic scale or the cosmological scale, new laws of physics (ones that we're not familiar with from everyday life) come to the forefront. So, we depend on calculations to tell us what to expect, and very often those mathematical models can be tested by doing experiments or building new telescopes. We may live in what seems like a bizarre, off-balance universe, filled with stuff like dark matter and dark energy, but is it really that strange? The stranger thing to me is that we, as humans, can actually conceive of these physical laws, ones that work at extraordinarily small and extremely large scales, and use these to interpret astronomical observations.

References

Abbott, B. P. et al. (2016), "Observation of gravitational waves from a binary black hole merger," *Physical Review Letters* 116, 061102.

Aghanim, N. et al. (2020) "Planck 2018 results," *Astronomy and Astrophysics*, 641, pp. A1–A12.

Akrami, Y. et al. (2020) "Planck 2018 results — VII. Isotropy and statistics of the CMB," *A&*, 641, A1.

The ATLAS Collaboration (2012) "Observation of a new particle in the search for the Standard Model Higgs boson with the ATLAS detector at the LHC," *Phys. Lett. B*, 716, pp. 1–29.

Brandenberger, R. H. (1985) "Cosmological perturbations in a universe dominated by a coherent scalar field," *Phys. Rev. D*, 32, pp. 501–503.

Dyson, F. W. and Eddington, A. S. (1920) "A determination of the deflection of light by the sun's gravitational field, from observations made at the total eclipse of May 29, 1919," *Philosophical Transactions of the Royal Society*, 220, pp. 291–333.

Guth, A. (2002) *The inflationary universe*, Available from: https://www.edge.org/conversation/alan_guth-the-inflationary-universe) (Copyright by Edge Foundation, Inc. All Rights Reserved.)

Einstein, A. (1915) "Die Feldgleichungen der Gravitation," *Sitzungsberichte*, pp. 844–847; (1916) "The foundation of the general theory of relativity," *Annalen der Physik*, 49, pp. 769–822.

Einstein, A. (1917) "Kosmologische Betrachtungen zur allgemeinen Relativitätstheorie," *Sitzungsberichte*, pp. 142–152.

Guth, A. H. (1981) "Inflationary universe: A possible solution to the horizon and flatness problems," *Phys. Rev. D*, 23, pp. 347–356.

Guth, A. H. (1997) *The inflationary universe*. Cambridge, MA: Perseus Books.

Heisenberg, W. (1927) "Über den anschaulichen Inhalt der quantentheoretischen Kinematik und Mechanik," *Zeitschrift für Physik*, 43, pp. 172–198.

Higgs, P. (1964) "Broken symmetries and the masses of gauge bosons," *Phys. Rev. Lett*, 13, pp. 508–509. See also Englert, F. and Brout, R. (1964) "Broken symmetry and the mass of gauge vector mesons," *Phys. Rev. Lett.*, 13, pp. 321–323.

Joule, J. P. (1845), *Philosophical magazine*, pp. 369–383.

Krauss, L. M. and Scherrer, R. J. (2007) "The return of a static universe and the end of cosmology," *General Relativity and Gravitation*, 39, pp. 1545–1550.

Linde, A. D. (1982) "A new inflationary universe scenario: A possible solution of the horizon, flatness, homogeneity, isotropy and primordial monopole problems," *Phys. Lett. B*, 108, pp. 389–393.

Linde, A. D. (1990) *Particle physics and inflationary cosmology*. Chur, Switzerland: Harwood.

Lykken, J. and Spiropulu, M. (2013) "The future of the Higgs boson," *Physics Today*, 66, pp. 28–33.

Markkanen, T., Rajantie, A., and Stopyra, S. (2018) *Front. Astron. Space Sci.* 5, pp. 40.

Misner, K. S., Thorne C. W., and Wheeler, J. A. (1973) *Gravitation*. New York: W. H. Freeman and Co.

Pasteur, L. (1882) "Mémoire sur les corpuscules organisés qui existent dans l'atmosphère: examen de la doctrine des générations spontanées," *Wellcome Collection*. wellcomecollection.org/works/njdg2696

Penzias, A. A. and Wilson, R. W. (1965) "A measurement of excess antenna temperature at 4080 Mc/s," *Astrophysical Journal*, 142, pp. 419–421.

Plionis, M. (2002) "The quest for cosmological parameters", *Lect. Notes Phys.*, 592, pp. 147–207. Available at: https://doi.org/10.1007/3-540-48025-0_7.

Pound, R. V. and Rebka, G. A. (1981) "Apparent weight of photons," *Phy. Rev. Lett.*, 4, pp. 337–341.

Albrecht, A. and Steinhardt, P. J. (1982) "Cosmology for grand unified theories with radiatively induced symmetry breaking," *Phys. Rev. Lett.*, 48, pp. 1220–1223.

Tryon, E. P. (1973) "Is the universe a vacuum fluctuation?", *Nature*, 246, pp. 396–397.

Whitaker, R. D. (1975) "An historical note on the conservation of mass," *Journal of Chemical Education*, 52, pp. 658.

Will, C. M. (2014) "The confrontation between general relativity and experiment," *Living Reviews of Relativity*, 17, pp. 4–117.

Chapter 2
Why Matter Doesn't Collapse

I do not mind if you think slowly, but I do object when you publish more quickly than you think.

—Wolfgang Pauli

Why doesn't all matter simply collapse down into a black hole? This may sound like a bizarre question, since clearly the Earth is solid under our feet. But when contemplating how to answer this, it's kind of amazing that matter doesn't collapse (since gravity pulls all matter together). Obviously, there must be some balancing force in atoms that prevents this from happening. The answer takes us into the realm of quantum mechanics and specifically to a principle first proposed a century ago by physicist Wolfgang Pauli.

First let's consider a simple atom like hydrogen. Our sun is made mostly of hydrogen gas, yet it also doesn't collapse under its own gravitation. We can look to hydrogen's properties to understand why this is so. Hydrogen is made from a single proton in its nucleus and a single electron "orbiting" the proton. (I use quotes here because, as you'll see below, the electron's motion is a bit more complicated than just circular motion.). You can't get much simpler in atomic physics than the hydrogen atom! The electron, with its negative charge, is attracted to the proton, with its positive charge, by the electric force. In hydrogen, both the electric force and the gravitational force between the electron and the proton are attractive. So, what keeps the electron in its orbit?

Here, it's tempting to make an analogy between the electron going around the proton compared with the Earth orbiting around our sun. The reason that Earth isn't sucked into the sun is because it has momentum (due to its velocity) and attraction from the sun's gravity pulls the Earth, causing its path to curve. A stable orbit occurs when there's a balance between the outward centrifugal force and the inward gravitational force. Does the same thing occur for the electron in a hydrogen atom? The answer is no, due to the weird properties of **quantum physics** (see Box 2.1). However, that didn't stop theoretical physicists from trying. What is described next shows the circuitous path that science often takes in getting to the correct answer.

Nature's Balancing Act. Ken Hicks, Oxford University Press. © Oxford University Press (2025).
DOI: 10.1093/9780197771471.003.0002

Box 2.1 A Brief History of Quantum Mechanics

The revolution in physics that took place in the early part of the twentieth century was rooted in experimental measurements that could not be explained with classical physics. For example, hydrogen gas emits only specific colors, seen as "lines," when the light is sent through a prism (which separates light into colors). Those lines came at regular spacings, and a mathematical model was found that showed the spacings were based on an integer relationship. The idea that whole numbers (1, 2, 3 . . .) had a role in atomic physics was unexpected. Almost all other measured quantities in physics were continuous, not quantized.

Soon, other integer relationships were discovered, such as a model by Max Planck to explain the spectrum (or colors) of light emitted by a hot object (called blackbody radiation) (Planck, 1900). For example, the burner on an electric stove will glow red at first; then, after getting hotter, it will glow orange. The spectrum of light from the sun also shows a spectrum consistent with blackbody radiation, the same as any other object heated to a high temperature. Planck's model used a sum over integers, assuming that energy was emitted in quantized amounts. So, integers show up again. Planck's formula also involved a new constant of nature, now called Planck's constant.

Planck's constant started to show up in other models of natural phenomena. Niels Bohr used it, along with integers, in his model of the hydrogen atom (see Box 2.2). Einstein used it to model the photoelectric effect, where light of a given frequency hits a metal surface, causing electricity to flow (Einstein, 1905). Einstein showed that Planck's constant was essential to calculate the smallest frequency of light for this to occur (for which he won the Nobel Prize). The point here is that Planck's constant was showing up again and again.

After these developments, Louis de Broglie showed that Bohr's assumption of quantization could be recast as saying the electron had wave-like properties (another Nobel Prize) (de Broglie, 1926). Soon after, Erwin Schrödinger published a paper showing that a wave equation for the electron could be used to get the same results for the hydrogen atom as Bohr, but this time the quantization came naturally in the mathematical solution (Schrödinger, 1926). That wave equation, now called Schrödinger's equation, also used Planck's constant, and is considered one of the seminal papers at the foundation of quantum mechanics. He won the Nobel Prize in 1933.

At about the same time, German physicist Werner Heisenberg published a paper using a new mathematical method (called matrix mechanics) that could also explain

continued

Box 2.1 *continued*

---

the spectrum of hydrogen, although it was slow to be accepted (Heisenberg, 1927). Later, it was shown that Heisenberg's formulation using matrix mechanics was equivalent to Schrödinger's equation. Heisenberg was awarded the 1932 Nobel Prize. Heisenberg's approach depends on a fundamental aspect of quantum theory, called Heisenberg's **uncertainty principle**, which says that there are fundamental limits to how well a measurement (such as position or momentum) can be done.

From these foundations, the quantum revolution took hold, and quantum mechanics was applied to almost all phenomena at the atomic and subatomic scale. Many other physicists contributed to the development of quantum theory. But suffice it to say that the success of quantum mechanics is unrivaled for small systems like atoms and that classical physics applies only to much larger objects.

In the early twentieth century, scientists (led by Niels Bohr) came up with a theory that would explain the hydrogen atom, assuming the electron goes in a circular orbit about the proton. In **Bohr's model** (Bohr, 1913), the electron has momentum, and the proton's positive charge pulls on the electron to curve its path, just like a planet orbiting the sun. The problem is that this "orbiting" theory doesn't hold up under closer inspection. It's known that electrons emit radiation (such as radio waves) when going around in a circle. The tighter the circle, the higher the energy of radiation. So, an electron in hydrogen, going around in a tiny circle, should emit radiation, reducing its momentum and spiraling inward until it meets the proton, essentially collapsing the hydrogen atom to a point. None of this made sense, since we know that hydrogen is stable and doesn't collapse in upon itself.

Bohr was forced to make an ad hoc adjustment to his model, putting in an unjustified criterion that the electron could only orbit at a discrete radius. That value of the radius was linked to a new constant of nature called **Planck's constant**. This constant had been discovered earlier by physicist Max Planck, solving another mystery of nature called **blackbody radiation** (more on this later). The use of Planck's constant in Bohr's model suggested that it played a fundamental role in the physics of atoms, even though the full theory (quantum mechanics) wasn't yet known to either Bohr or Planck. Even though Bohr's solution worked (his model could explain some properties of hydrogen, see Box 2.2), it had this unjustified assumption. So, they still had a problem: what keeps the electron in orbit?

Box 2.2 Niels Bohr

Niels Bohr is one of the fathers of quantum mechanics. Born in Denmark in 1885, he got his PhD in physics but was hired as a professor of physiology at the University of Copenhagen in 1912. Just prior to taking that position, he developed the foundations of his theory of atoms during a fellowship in England working with Ernest Rutherford and others. His papers on the hydrogen atom (Bohr, 1913) catapulted him to fame (and a Nobel Prize).

The origins of the planetary model of the hydrogen atom, where the electron makes a circular orbit around its nucleus (a single proton), are rooted in the colors of light that are emitted by hydrogen under electric discharge. This "lightning in a bottle" happens when any gas is captured in a glass container and high voltage is applied to either side. The resulting spectrum is not what you might expect. Instead of a continuous range of colors being produced, only narrow "lines" of color are seen. (These can be seen by passing the light through a prism, for example.) The extraordinary thing about the distance between the lines is that they can be described by a simple mathematical equation that involves only integers.

Having an integer relationship in a physical process is not common. For example, if you measure the velocity of a ball, it can have any speed, not just integer multiples of some fundamental speed unit (such as 1 or 2 or 3 units, but not 1.5 units). Similarly, the spectrum of light from the sun shows a continuous range of colors, not just individual lines of color. The fact that integers show up suggests that the hydrogen atom has a simple structure (such as circular orbits of the electron).

A decade earlier, German physicist Max Planck had published a paper showing that the intensity of colors coming from any hot object (be it the sun or red-hot metal) can be described if he assumed that the wavelengths of light are quantized (see Box 2.1). Bohr used Planck's constant to quantize the electron's orbit (specifically its angular momentum) and found that, with this single assumption, he could describe the lines of color emitted by hydrogen. The near-exact agreement between Bohr's model of the hydrogen atom and Nature caused a major stir in the scientific world.

Although today we understand that the hydrogen atom is a bit more complex than the simple picture of Bohr's model, the quantization seen by Planck, Einstein, and Bohr were the birth of new ideas that soon led to the theory of quantum mechanics. For years after this, Bohr was revered as one of the luminaries in the field of quantum science. For example, when Einstein questioned the basis of quantum mechanics, he had public debates with Bohr, where Einstein quipped "God does not play dice" in

continued

Box 2.2 *continued*

---

reference to the co-called Copenhagen interpretation of quantum phenomena with probability and uncertainty as its core idea.

For a more entertaining insight into Bohr and his role in Europe during World War II, see the play *Copenhagen* by playwright Michael Frayn. Bohr also makes a cameo appearance in the blockbuster movie *Oppenheimer*. These are just some examples of Bohr's lasting influence in the history of physics.

The answer wasn't understood until the equations of quantum mechanics were formulated during the 1920s. A fundamental assumption of quantum systems is that there is a minimum amount, or "quantum," of energy that can be exchanged between two objects. Furthermore, all energy in atoms come in integer quantities of this minimum amount (i.e., energy comes in quantized units). This idea became a postulate of quantum mechanics. It's called a postulate because it can't be proven and must be assumed true. For example, one of the postulates of geometry is that a straight line is the shortest distance between two points. Once this postulate is assumed true, then many theorems of geometry follow. Similarly, once the postulates of quantum mechanics are assumed, then it results in many theorems that explain a range of natural phenomena, such as the stability of the hydrogen atom.

So, quantum mechanics says that an electron in a hydrogen atom has a minimum energy and can't emit that energy as radiation. That minimum energy corresponds to a given momentum of the electron, and remarkably it exactly matches the momentum used in Bohr's model, where a discrete radius was assumed. However, quantum mechanics went beyond this and made other predictions about the properties of the hydrogen atom that Bohr's model failed to explain. So, quantum mechanics could explain the stability of the hydrogen atom using just a few simple postulates. In addition, the mathematics of quantum mechanics could be used to predict a host of other things, such as radioactive decay.

It might sound like circular logic, where nature tells us that hydrogen is stable and so we invent a set of equations that forces hydrogen to have a minimum energy so that it's stable. But it's more than that. Once we have the equations of quantum mechanics, we can apply them to all sorts of other situations. For example, we can predict what wavelengths (colors) of light that are emitted by hydrogen gas in a hot environment (like in the sun). That color spectrum matches exactly with what's measured for the sun. Going further, we can predict the properties of other atoms, like carbon or oxygen, and describe things such as chemical bonding. All of this stems from the equations and the postulates of quantum mechanics.

Today, an understanding of quantum mechanics is used in the development of all modern electronics. The transistor, which is at the heart of all computers, was discovered by applying the rules of quantum mechanics to semiconductor materials such as silicon. Furthermore, all subatomic systems obey the laws of quantum mechanics, so nuclear reactions that go on in our sun (or inside a nuclear power plant) are understood by application of these equations. The importance of the quantum revolution can't be understated. Even the simple fact that all matter doesn't collapse in upon itself is a consequence of the laws of quantum mechanics, as discussed next.

2.1 The Pauli Exclusion Principle

One of the assumptions of the quantum revolution, which was inexplicable when it was first proposed, is called the **Pauli exclusion principle** (named after the physicist Wolfgang Pauli who proposed it back in 1925, see Box 2.3). This principle states, in its simplest form, that no two electrons can occupy the same quantum state of a system. Because there's a lot packed into that principle, before going further it's important to know what is meant by a "quantum state."

Box 2.3 Wolfgang Pauli

One of the most colorful characters in the development of quantum theory is Wolfgang Pauli. He was born in Vienna and educated at the University of Munich, receiving his PhD in 1921. Like many of the great theoretical physicists of the time, he studied for a time with Niels Bohr.

Pauli was a perfectionist and a critic. He would publish a paper only when he was sure that he could explain it clearly, and he demanded that others do the same. (See the quote by Pauli at the start of this chapter.) Pauli didn't hesitate to let his feelings be known to any poor physicist who happened to present their ideas without solid footing. However, at times Pauli had a sense of humor and always a quick wit. He had a warmer side when among friends, even if he was brutally honest when discussing science. Many humorous quotes by Pauli can be found in biographies of this great man.

continued

Box 2.3 *continued*

Although he published few papers, he had a big impact on the development of quantum mechanics through letters he wrote to colleagues. His research was centered on atoms with many electrons, showing that electrons are collected in "shells" (like layers of an onion, one group inside the other). The properties of chemistry, embodied by the periodic table of elements, are largely dictated by the number of electrons in the outer shell of an atom.

To explain the shell structure of atoms, Pauli introduced a new quantum number that took on only two values (either "up" or "down") and is now known to correspond to spin. To get the shell structure of atoms, Pauli formulated his exclusion principle, saying that no two electrons can exist having the same quantum numbers. For example, two electrons can exist in the innermost shell of an atom, one electron in the up state and one down.

In addition to the exclusion principle, Pauli introduced the idea of a new particle, the neutrino, to solve a problem with conservation of energy seen in radioactive decay (Pauli, 1930). It took many decades for technology to advance to the point where a neutrino could be detected, showing that Pauli was decades ahead of his time. Today, the neutrino is one of the most important building blocks of the Standard Model of particle physics.

Another pillar of theoretical physics, called the spin-statistics theorem (Pauli 1940), is also attributed to Pauli (based on earlier work by M. Fierz). This theorem separates particles into two groups: those with integer spin (called bosons) and those with half-integer spins (called fermions). This distinction into two classes of particles has important implications (see the main text).

Any one of the above contributions would have landed Pauli's name in the textbooks. The fact that these (and other) results came from a single individual is truly remarkable.

In the theory of quantum mechanics, it's possible to solve the equations such that a quantum system, say the hydrogen atom, can only have integer values of a quantity, like energy. If we denote the minimum energy by E, then a simplified picture of a quantum system, it can jump to energy $2E$ or $3E$ (and so on) but not any arbitrary energy in between. This is very different from our everyday experience, where measuring the energy of a moving ball can take on any value, from zero to a large value (and seemingly any number in between). In a quantum system, like the electron in hydrogen, the electron is always moving with some minimum amount of energy. The quantum state of a system is specified by the integer values that come from solutions of the quantum equations. For example, a quantum system might be in "state 1," corresponding to the minimum energy E, or "state 2" if it has gained energy up to $2E$. Similarly, if the system starts in state $3E$, as shown in Figure 2.1, it can transition to state $2E$ at a larger radius and give off a **quantum of energy**, hf, by emitting a photon. Here, h is Planck's constant, and f is the frequency of the light emitted.

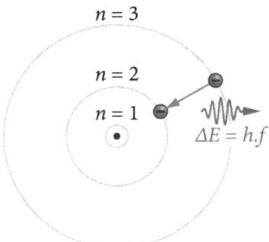

Figure 2.1 Schematic diagram of the transition in a hydrogen atom of an electron from the quantum state at 3E going to the 2E state, by emitting a photon with energy $\Delta E = hf$. The integer given by n is the principal quantum number derived from quantum mechanics.

So, the **quantum state** of a system is specified by the "quantum numbers" of the system, which refers to the integers that appear naturally from solving the quantum equations. In the case of hydrogen, several integers are needed to specify its quantum state. In addition to the principal quantum number, n, that specifies the energy, there is another integer that specifies the orbital motion of the electron and yet another integer that tells about how the atom behaves inside a magnetic field. Those who have taken a chemistry class may be familiar with these integers, because these quantum numbers are used to categorize atoms on the periodic table. In other words, the chemistry of elements depends on the quantum states of electrons in these atoms.

Getting back to the Pauli exclusion principle, the fact that no two electrons can be in the same quantum state is fundamental to the structure of matter. It is the reason that the electrons in heavier elements don't all have the same minimum energy (i.e., the same principal quantum number). An electron that is excluded from "state 1" must move up to "state 2," where its quantum numbers are different. As a result, electrons form shells, see Box 2.3, one electron orbit inside another. In detail, the picture is more complicated, because the electron orbits aren't all circular (the shape and size of the orbits change with quantum numbers) and also because electrons have a property called "spin" (see below). But the central idea to take away is that the complex structure of atoms, which underlies all of chemistry and biology, depends on both quantum mechanics and the Pauli exclusion principle.

What is the physical basis of Pauli's exclusion principle? Why is Nature so contrived as to require that electrons be unfriendly, always excluding other electrons from their territory? Well, there is no simple explanation. Any attempt to explain it always comes down to making another postulate that's mathematically equivalent. But without the exclusion principle our world would clearly be very different. The fact that the Earth is solid under our feet is due to this exclusion principle.

So, this chapter started with a speculation that there must be a balance between the gravitational forces in the sun, pulling it inward, and some other force. Although it's not called a "force" by physicists, the exclusion principle is simply a fact of Nature, where electrons are kept apart so that they don't occupy the same quantum state. It's such a simple principle but has such vast consequences. The stars, the planets, and life itself depends on the Pauli exclusion principle. Nature, for whatever reason, has provided an offset between the physical forces that pull matter together and the exclusion principle that pushes it apart. Perhaps someday in the future scientists will understand the exclusion principle from a more fundamental theory (like string theory), but for now we just assume it to be true.

What would the universe look like if Pauli's exclusion principle were not true? Then all matter would collapse in upon itself. After the Big Bang, matter would have gathered together, making a bunch of black holes, eventually merging into a huge single black hole. In short, the universe would not exist. Yet this important principle is not really understood. It is assumed, just like the postulates of quantum mechanics. As much progress as science has made, this example shows how fragile our understanding of Nature is.

2.2 The Spin Quantum Number

The electron is categorized among a class of particles that are called **fermions** (named after the physicist Enrico Fermi, a pioneer of nuclear physics). All known particles can be divided into two groups—fermions and **bosons** (the latter group named after Indian theoretical physicist Satyendra Bose)—based on an intrinsic property of the particle called **spin**. As we shall see, depending on whether a particle is classified into one group or the other makes a big difference to how it behaves, such as whether it obeys the Pauli exclusion principle or not.

The word "spin" is a bit misleading because we can't actually build a microscope sufficiently powerful to see a particle's rotation. The term originates from envisioning particles as tiny spheres, with the particle spinning on its axis. From known laws of physics (Maxwell's equations), a spinning sphere having an overall electric charge creates a magnetic field, much like the spinning Earth has a magnetic north pole and south pole. The Earth, we believe, has electric currents flowing deep in its interior, causing a weak magnetic field at the surface that can be measured with a compass.

From laboratory experiments, we know that the electron behaves as if it carries a tiny bar magnet inside. In other words, the electron has a north pole and a south pole. We can measure the effect of an electron in a laboratory magnetic field, such as an electromagnet. When the magnet is off, the electron's north pole can point in any random direction. When the magnet is turned on, however, the electron tries to orient itself to align with the magnetic field, just like a compass needle will align itself with the direction of Earth's magnetic field. If the electron is oriented the wrong way, its spin direction will flip, causing a small amount of energy to be released, which can be measured. Similarly, if a compass needle is oriented opposite to the Earth's magnetic field, it will flip its direction.

Another way to see the effect of the electron's magnetic property is to put a tube of hydrogen gas into a magnetic field and run an electric current through it. The hydrogen gas will emit specific colors (wavelengths of light), which can be seen by viewing it through a prism (or using a more compact device such as a diffraction grating, which also spreads out light into a color spectrum). With the magnet off, the colors appear as sharp lines. When the magnet is turned on, some of the colored lines split into several isolated lines when magnified. This is called the Zeeman effect (named after the scientist who discovered this phenomena) and is further evidence that the electron has spin.

Even though it's unlikely that the electron is actually spinning, the terminology is now pervasive, so let's just use "spin" to represent the magnetic property of a particle. Most fundamental particles have spin, and this is one of their quantum numbers. For example, protons and neutrons also have spin and, just like the electron, are classified as fermions. These particles have spin "half," meaning that their spin quantum number can either be $+1/2$ or $-1/2$ (meaning aligned or anti-aligned with a magnetic field) but never zero as shown in Figure 2.2. The transition from $-1/2$ to $+1/2$ is exactly one unit of spin, which is measured in units of Planck's constant.

The other classification, particles called bosons, have integer values of spin, such as spin-1 or spin-0. It may seem like a fine distinction, but Nature can

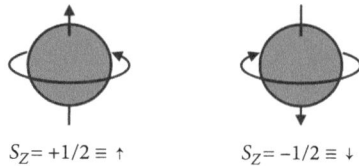

$S_Z = +1/2 \equiv \uparrow$ $S_Z = -1/2 \equiv \downarrow$

Figure 2.2 Schematic of a fermion particle, having spin quantum number 1/2, corresponding to spin "up" (left) and spin "down" (right). We don't know whether particles are really spinning, and this picture is only to help visualize the measurable magnetic property of fermions.

Adapted from Maschen (2011). Via Wikimedia Commons (public domain).

be subtle. It turns out that only particles with half-integer spin, fermions, obey the Pauli exclusion principle, whereas bosons are not restricted in this way. This makes a huge difference to how the particles behave. In particular, bosons can cluster together, all having the same quantum numbers. For example, the photon (a quantum of light) has spin-1, making it a boson. Two photons can pass right through each other, like two waves on the surface of the ocean. In contrast, two electrons will collide, bouncing off each other like two beach balls on the water.

The **Higgs boson**, discussed in Chapter 1, is the only fundamental particle we know that has spin-0. This means that a Higgs boson (see Box 2.4) can "pop" out of the vacuum as a quantum fluctuation. Other particles have non-zero spin and can only be created (from pure energy) as pairs of particles, so that the law of angular momentum is conserved (spin is one form of angular momentum, which is the momentum of rotating things). The quantum fluctuation that started the Big Bang, in the inflation scenario, is thought to be a particle with spin-0, but heavier than the Higgs boson.

Box 2.4 Peter Higgs

Peter Higgs can be described as humble and a classic English gentleman. He also made one of the most spectacular theoretical predictions of the past century, for a new type of massive particle (the Higgs boson) that carries his name. The importance of this particle, now known to exist, is that it forms a cornerstone of the Standard Model of particle physics (see Chapter 4).

Higgs came upon the idea when trying to solve an obscure problem in the theory of fields (for example, the electric field that surrounds all charged particles). At the time, this subject was primarily of interest to only a few theoretical physicists.

The problem comes while trying to explain the properties of radioactive decay, which can be modeled using equations for the exchange of very massive particles (the W and Z bosons).

Physicists had known for a long time (since about 1940) that the strong-nuclear force could be described mathematically by the exchange of particles, and the weak-nuclear force was similar except that the exchange particles have a very large mass. In contrast, the electromagnetic force can be modeled by the exchange of a massless particle, the photon. To change the field equations to get a massive exchange particle, the math predicted some fictional massless particles (called Goldstone bosons) that were not seen in Nature. To get rid of these fictional particles, Higgs introduced a new way to cancel those math terms in the equations but resulted a new term corresponding to a massive boson with zero spin (Higgs, 1964).

All of this may seem rather esoteric, and so it seemed at the time, even to Peter Higgs. It turns out that the original version of Higgs's paper did not mention the massive boson; he wrote the paper with the goal of solving a dilemma in field theory. He only revised his paper to mention this after receiving comments from the journal's referee who reviewed (and rejected!) the original version (Edinburgh, 2014). However, the results in this paper were important and soon used elsewhere in physics (see Chapter 1). Soon, the Higgs mechanism was an indispensable part of the theoretical physics toolchest.

In an interview, Peter Higgs is quoted as saying that the discovery of the Higgs boson "ruined my life" (Close, 2022). This gives some insight into his personality. He went on to say, "My relatively peaceful existence was ending. My style is to work in isolation and occasionally have a bright idea." Not all physicists are vying for the Nobel Prize. Some are doing it for the sheer love of the mathematics and the understanding how Nature can be described by equations. For the many scientists who are doing what they do without the recognition of a prize, you can take heart from the words of Peter Higgs.

It should now be obvious that it's a big deal to Nature as to whether a subatomic particle is a fermion or a boson. If the electron were a boson, then electrons wouldn't form shells in atoms, and the Earth wouldn't be solid. The stuff of ordinary matter (electrons, protons, and neutrons) is made from fermions, and the particles that exchange energy (such as photons and gluons, which bind the nucleus together) are bosons. The quantum property of particles that we call spin is very important to providing a balance to nature.

2.3 Where Does Spin Come From?

Experimental evidence of the electron's spin was first seen in a famous experiment by **Stern and Gerlach** (Gerlach and Stern,1922). At the time, this was a complete surprise to physicists, and there was no theoretical prediction of this phenomena. Wolfgang Pauli proposed his exclusion principle a few years later (Pauli, 1925), even though he didn't understand why the electron had a spin. It wasn't until years later, when theoretical physicist Paul Dirac formulated an equation (Dirac, 1928) that merged the ideas of quantum mechanics and Einstein's special relativity, that the spin quantum number could be understood mathematically.

Dirac's equation applies only to particles having a spin of 1/2 (in other words, to fermions). This is because the structure of Dirac's equation, which uses matrix algebra, allows the transition of the particle between two definite quantum states. This is exactly what is needed to describe the results of experiments (such as the one by Stern and Gerlach) where the electron can have two states: spin "up" (aligned with a magnetic field) or spin "down." This solution of Dirac's equation comes out directly due to its mathematical form.

Put simply, the spin quantum number arises naturally from the terms in Dirac's equations. Those equations can be used to predict measured quantities, such as how an atom behaves when a magnetic field is applied. The key point is that Dirac's equation is consistent with Einstein's theory of special relativity. The original (nonrelativistic) formulation of quantum mechanics, embodied by **Schrödinger's equation** (see Box 2.5), doesn't have the form required by relativity. As a result, there are no terms in Schrödinger's equation corresponding to spin, even though spin can be added in (without justification) as an ad hoc "fix" to Schrödinger's equation.

Box 2.5 Erwin Schrödinger

Perhaps the most famous among the founders of quantum mechanics is Erwin Schrödinger. Born in Vienna, where he studied as a student, he entered academia and eventually landed as a full professor at the University of Zurich, where he did the seminal work for Schrödinger's equation that is a staple in all textbooks on quantum mechanics.

The journal paper where he presented the Schrödinger wave equation (Schrödinger, 1926) is considered one of the most important physics publications of the twentieth century. Like Heisenberg's uncertainty principle, it leads to calculations that give only probabilities for a measurement, not certainty. Schrödinger himself was never quite comfortable with the probability aspects of quantum mechanics (called the Copenhagen interpretation, see Box 2.2) and wrote, "I don't like it, and I'm sorry I ever had anything to do with it."

To illustrate this point, Schrödinger provided a controversial (and oft-quoted in popular media) thought experiment called the Schrödinger's cat paradox. In this thought experiment, he has a cat in a closed box that could be either alive or dead, depending on the outcome of a quantum measurement (such as radioactive decay). If the probability interpretation is correct, then one can only calculate using Schrödinger's equation a probability for the outcome (i.e., whether the cat is alive or dead). He used this example to point out the ridiculousness of this interpretation: a cat can't be 50% alive and 50% dead; it must be one or the other!

The probability interpretation of quantum mechanics continues to confuse people to this day. However, there seems to be no alternative. Quantum mechanics has always triumphed in comparison to subatomic measurements, and no competing theory has such predictive power. All attempts to overcome the probability interpretation have failed, and interested readers can look up Bell's theorem for more details. It seems that we are stuck with the uncertainty of how subatomic systems behave and our current formulation of quantum mechanics.

Schrödinger's life is not without controversy. He suffered from tuberculosis in the 1920s and went to a sanatorium in Switzerland several times. Although married, he had various affairs with women, some of them quite young according to a journal he kept (Humphreys, 2021). For this behavior, his portrait has been removed from some academic institutions. He also had an interest in philosophy and was fascinated by the connection between consciousness and Nature. Some of his writings on this subject have raised eyebrows among scientists, such as saying "Consciousness cannot be accounted for in physical terms" (Schrödinger, 1958). Whether you agree with this or not, it shows that he was deeply concerned with how the human mind works.

Regardless of what people may think of his personal life, Schrödinger's equation was an achievement worthy of the Nobel Prize he received. It was a brilliant insight into how physics at the subatomic scale is different from that of our daily world, even if the probabilistic interpretation makes one question whether a deeper theory of Nature exists.

If you think about it, Dirac's equation is rather remarkable. What should special relativity have to do with a particle's spin property? Special relativity is all about the relationship between matter and energy for moving particles, resulting in equations like $E=mc^2$. However, a particle has its spin property even if the particle isn't moving (i.e., has zero velocity). Yet Dirac found that combining the requirements of special relativity with the ideas of quantum mechanics results in an equation that predicts the behavior of fermions. Coupling this with the above fact that fermions obey the Pauli exclusion principle (and hence results in matter not collapsing in upon itself), we see that it's a very important development to the understanding of Nature's balancing act.

There is another remarkable feature of Dirac's equation that wasn't understood for many years. Although it has nothing to do with spin, Dirac's equation has an additional two quantum states in its solution, for a total of four quantum states (the two for spin and the two new states). The two new states appear to have "negative energy," which seems ridiculous since the particle's mass can't be negative and nor can it have negative kinetic energy. In physics, energy is *defined* as a positive number, and so these other two quantum states from Dirac's equation appeared to be nonsense. However, today, we understand these quantum states to correspond to **antiparticles** (or "antimatter"). This topic is covered in Chapter 3, but it's amazing that the concept of both spin and antiparticles comes about naturally from Dirac's equation.

Just to complete the picture, a different equation, called the **Klein-Gordon equation**, combines special relativity and quantum mechanics for bosons (particles with integer spin, which don't obey the Pauli exclusion principle). The Klein-Gordon equation (Klein, 1926; Gordon, 1926) also has negative energy solutions. So, antiparticles arise in the equations for both fermions and bosons, but only when the requirements of special relativity are combined with the ideas of quantum mechanics. The fact that antiparticles exist can be shown from experiments where high-energy particles collide, resulting in a spray of new particles coming from the collision point (some being antimatter particles). Antiparticles are also very important to the evolution of the universe. More about this in Chapter 3.

2.4 Supersymmetry and Grand Unified Theory

Why should some particles be fermions and others be bosons? This is a question that has mystified physicists for many years. To answer this question, some theorists have speculated that there could be some kind of connection between fermions and bosons. One proposed idea, called **supersymmetry**, is that for every type of fermion, there is a corresponding boson, and vice versa. If the theory of supersymmetry is correct, it would provide a kind of balance between bosons and fermions that might explain why both types of particles are needed.

The motivation for the supersymmetry concept is difficult to explain without equations, but the main idea is that for each fermion there is a corresponding boson (and vice versa). It might seem odd to think that particles with half-integer spin are somehow connected with particles having integer spin (where is the symmetry?), but this is needed to solve a theoretical dilemma. When a theoretical physicist views empty space, quantum mechanics says that particles (and their antiparticles) are constantly popping in and out of existence on very short timescales. This comes from the uncertainty principle, which says that the energy of empty space is not zero because the energy is uncertain (over a very short timescale). So, over a short period of time such as 10^{-25} seconds (much too short for humans to notice), particles can be created and destroyed in empty space. This happens all over the universe and would result in a huge amount of extra mass. However, if supersymmetry is correct, there is a cancellation in the equations (fermions and bosons have terms with opposite signs), and this spurious mass goes away! Without supersymmetry, this theoretical dilemma is difficult to resolve.

You might consider speculations about empty space having excess energy as esoteric and maybe even pointless. However, if we are to trust the equations of quantum physics, which do remarkably well in predicting laboratory measurements, then we need to take seriously the implications of the uncertainty principle. In that case, there is a dilemma that either empty space contributes to the mass of the universe (going against the standard big bang model) or something else like supersymmetry exists in nature. Whatever the explanation, be it a model like string theory or something else, this is an unsolved problem in theoretical physics.

There is no experimental evidence yet for supersymmetry (also called SUSY for short), but theoretical physicists suggest that the SUSY particles (Kaku, 1993) could exist only at high energies (with large mass) and are not seen in current lower-energy experiments. One of the motivations to build larger accelerator facilities, such as the **Large Hadron Collider** (LHC) located in Switzerland, is to collide protons at higher energies to see whether new large-mass particles can be created. So far, the only new large-mass particle discovered by the LHC is the Higgs boson, which was expected based on the Standard Model of particle physics (and not part of the new SUSY particles). But still the theorists persist, suggesting that the SUSY particles exist at higher energies than can be achieved at the LHC, perhaps only at energies that were present at the start of the Big Bang.

Another reason that theorists supporting the SUSY model are persistent is because this symmetry would solve a problem in particle physics called the **hierarchy problem**. In short, the nuclear force is vastly stronger than the gravitational force, making nuclear the "king" of forces. If, at the first instances of the Big Bang, all forces were created equal (meaning that they all acted in nearly equal measure at this super-high temperature), then there is a chance to unify all forces under a single equation.

It has been the dream of many theoretical physicists, including Einstein, that all the forces of Nature can be unified into a single theory, called a **Grand Unified Theory** (GUT). To explain this, let's first look at some history.

In the past, advances in theoretical physics have shown that some seemingly unrelated forces could be unified. For example, you might think that the electric force between two charged objects is not related to the magnetic force. Iron can be magnetized (by putting it close to another magnet) and have no net charge. Similarly, an object such as a balloon can be given excess charge (by rubbing it across material like a wool sweater), and it has no magnetic field. But even though the electric and magnetic forces appear different, the results of experiments and the subsequent theoretical work by James Clerk Maxwell (a Scottish scientist from the 1800s) showed that these two forces were unified into one set of equations. We now talk about a single force, the electromagnetic force, that unifies these two phenomena.

Similarly, in the late 1950s, a theoretical model was developed to unify the electromagnetic force and the weak-nuclear force. The latter is responsible for radioactive decay, where a nucleus such as potassium can spit out an electron, turning a neutron into a proton inside the nucleus. Again, these seem like entirely different forces. The electromagnetic force doesn't require radioactivity, and radioactivity happens only to subatomic particles at small distances, whereas electromagnetic waves (like radio waves) can be sent over long distances. Yet there is a connection between the two, and it took many decades before it became clear that the theoretical model of a unified "electroweak" force was correct. The reason we know this is because predictions of this model were tested again and again by experiments, with the result that the electroweak theory could predict the measurements with precision.

Having unified electromagnetism and the weak-nuclear force, theoretical physics was on a roll. Could one also unify the electroweak theory with the strong-nuclear force? The latter is responsible for holding the nucleus together, essentially the attractive force between neutrons and protons. Even more enthusiastic theorists wanted to marry gravity into the unified-force family. Although there is not yet a successful theory to do this, many physicists believe that this will happen. More theoretical work is needed, and if SUSY particles are discovered from building higher-energy accelerators, that would give theorists the empirical evidence they need to guide development of new mathematical models.

While the weak-nuclear and electromagnetic forces appear very different at ordinary energies (the energies we encounter in daily life), one can show that these forces have about the same "strength" (the amount of force needed to pull two objects apart) when experiments are done at high-energy accelerators, like the LHC. Theorists can estimate the energy scale needed before the electroweak and strong-nuclear forces are to become unified. It turns out to be a very high energy, well above any accelerator we could build today. This brings us back to the hierarchy problem and supersymmetry.

Since the energy scale needed to unify the known forces is above any energy we can reach on Earth but did exist in the first moments of the Big Bang, we turn to cosmology for answers. Our universe may exhibit some remnants of the once-unified forces. There may have been a symmetry between fermions and bosons present at the start of the Big Bang, and the subsequent breaking of this symmetry (that occurred as the universe expanded) resulted in giving us the fermions and bosons we see today. If such a symmetry existed, and SUSY is only one such theoretical model, then this could explain why quantum corrections (also called quantum loops) are needed to get the forces to unify, as desired by the GUTs.

The reason such a high energy is needed is because of the hierarchy problem (Arkani-Hamed, Dimopoulos, Dvali, 1998). There is an ordering to the forces of nature, with the strong-nuclear force being the strongest, followed by the electromagnetic force (about 1% of the strong force when inside the nucleus), then the weak-nuclear force (at about 0.1% weaker than the electromagnetic force), with gravity coming in by far as the weakest. In fact, gravity is about forty orders of magnitude weaker (10^{-40} weaker) than the strong-nuclear force. Why should gravity be so much weaker than the other forces? The answer is not known, and that's why it's called the hierarchy problem. But to unify the strength of these forces, it takes about forty orders of magnitude in the energy scale to achieve unification with gravity. Building an accelerator that is only one order of magnitude (i.e., one factor of ten) higher in energy is costly, and so increasing the energy by many orders of magnitude is out of the question. This is why such enormous energy is needed to succeed with GUT models. But even if we can't reach the GUT energy scale, there may be remnants of the GUT predictions at lower energies.

Whether any of this theoretical speculation is true or not, only measurement can verify it. Hence, SUSY has made estimates (not yet firm predictions, because there are uncertainties in the parameters of the theory) of possible new particles that could be produced by accelerators such as the LHC. It appears that those original estimates were off, and if any SUSY particles exist, they are beyond the reach of current experiments. But without the SUSY particles, and the quantum corrections that would allow theorists to get to the GUT energy scale, it will be much harder to achieve the dream of grand unification of all known forces of nature.

2.5 Back to Nature

Getting back to the real world, Nature has a mixture of particles, some fermions and some bosons. We don't know why this is, and theorists are searching for ideas of a more fundamental theory that could explain it. Such ideas of theoretical unification have worked in the past, but it may be many years before physicists hit on the correct mathematical equations, if any exist, to explain why we have these two classifications of particles.

One class of particles, fermions, obey Pauli's exclusion principle. The origin of this principle is not known, but perhaps it will be explained by future theoretical research. Regardless, this principle is a bedrock of quantum mechanics. Without it, our universe would not exist.

You might be left with the impression that there are lots of things we can't explain about Nature. Indeed, this is correct. But we have come a long way in understanding the world around us in the past hundred years since quantum mechanics and general relativity were first formulated. There are still a lot of fundamental questions to answer about Nature, and this is a good thing for students who want to make an impact in the field. There are many future Nobel Prizes left for those who can find equations that explain some of the mysteries of particle physics.

References

Arkani–Hamed, N., Dimopoulos, S., and Dvali, G., (1998) "The hierarchy problem and new dimensions at a millimeter," *Physics Letters B*, 429, pp. 263–272.

Bohr, N. (1913) "On the constitution of atoms and molecules," *The London, Edinburgh, and Dublin Philosophical Magazine and Journal of Science*, 26, pp. 1–25.

Close, F. (2022) *Elusive: How Peter Higgs solved the mystery of mass*. New York: Basic Books.

de Broglie, L.-V. (1926) *Ondes et mouvements*. 1926 Solvay Conference.

Dirac, P. A. M. (1928) "The quantum theory of the electron," *Proceedings of the Royal Society A*, 126, pp. 360–365.

Edinburgh, School of Physics and Astronomy (2014) *Brief History of the Higgs Mechanism*. Available at: https://www.ph.ed.ac.uk/higgs/brief-history.

Einstein, A. (1905) "On a Heuristic Viewpoint Concerning the Production and Transformation of Light," *Annalen der Physik*, 17, pp. 132–148.

Gerlach, W. and Stern, O. (1922) "Der experimentelle Nachweis der Richtungsquantelung im Magnetfield," *Zeitschrift für Physik*, 9, pp. 349–352.

Gordon, W. (1926) "Der Comptoneffekt nach der Schrödingerschen Theorie," *Zeitschrift für Physik*, 40, pp. 117.

Higgs, P. W. (1964) "Broken symmetries and the masses of gauge bosons," *Phys. Rev. Lett.*, 13, pp. 508–509.

Humphreys, J. (2021) "How Erwin Schrödinger indulged his 'Lolita complex' in Ireland," *The Irish Times*, December 11.

Kaku, M. (1993) *Quantum field theory*. New York: Oxford University Press, p. 663.

Klein, O. (1926) "Quantentheorie und fünfdimensionale Relativitätstheorie," *Zeitschrift für Physik*, 37, pp. 895.

Pauli, W. (1925) "Über den Zusammenhang des Abschlusses der Elektronengruppen im Atom mit der Komplexstruktur der Spektren," *Zeitschrift für Physik Physik*, 31, pp. 765–783.

Pauli, W. (1930) "Letter to the Tübingen meeting, dated Dec. 4, 1930," Pauli Archive, CERN.

Pauli, W. (1940) "The connection between spin and statistics," *Physical Review*, 58, pp. 716–722.

Planck, M. (1900) "On an Improvement of Wien's Equation for the Spectrum," *Verhandlungen der Deutschen Physikalischen Gesellschaft*, 2, pp. 202–204.

Schrödinger, E. (1926) "An undulatory theory of the mechanics of atoms and molecules," *Physical Review*, 28, pp. 1049–1070.

Schrödinger, E. (1958) "In Collected Papers, Vol. 4," Austrian Academy of Sciences, Vieweg & Sohn, p. 334.

Heisenberg, W. (1927) "Über den anschaulichen Inhalt der quantentheoretischen Kinematik und Mechanik," *Zeitschr. Phys.* 43, pp. 172–198.

Chapter 3
Mirror, Mirror in the Sky

A theory with mathematical beauty is more likely to be correct than an ugly one that fits some experimental data.

—P. A. M. Dirac

Years ago, I was advisor for a doctoral student who is also a champion swimmer. Before he became a graduate student, he trained for a spot on the US Olympic team. In the race to determine who would go to the Olympics, he missed the cutoff by only 0.01 second. You can imagine his disappointment! But he had trust in the accuracy of the timer. Digital timers have been in common use now for over fifty years, and typically have a precision that is better than 0.0001 second. If you're going to miss a chance to go to the Olympics, you better know that the timing mechanism is fair and precise.

Precision and accuracy in scientific measurements is essential. For example, consider an old-style balance used to measure weights. I envision two platforms on either side connected by a rigid beam that rests on a knife edge. How precise do you suppose it is? If you had a balance like this set up in your house, could you get a 1% precision? Getting to that precision, one part in a hundred, should be easy, assuming you have a set of standard weights that came with the balance. (The standard weights are the guarantee of accuracy—getting the correct value—which is different from precision.) Could you reach a precision of one part in 1,000? How about one part in 1,000,000? The latter precision would be extremely difficult to get with a tabletop balance!

For the remainder of this chapter, the topic is about a small imbalance in the laws of physics that are only about one part in ten billion. To show this, experiments must be carried out with incredible precision. Knowing the level of precision that can be reached by your instrument is obviously crucial to knowing whether you can trust it.

When scientists report on a measurement (e.g., some quantity, such as the cosmic microwave background of the Big Bang), the paper will include some statement about the precision of the measurement. This information is often lost in media reports, which are usually written for the general public. But the measurement uncertainty is necessary for the peer-review process, which is

Nature's Balancing Act. Ken Hicks, Oxford University Press. © Oxford University Press (2025).
DOI: 10.1093/9780197771471.003.0003

used by most scientific journals to verify the precision of the results. It is a critical part of the scientific process.

For example, we know[1] that the composition of the universe is about 68% from dark energy, 27% from dark matter, and 5% from ordinary matter. The truth is that one doesn't measure these percentages directly—they are valid only within the context of a theoretical model that is used to interpret a variety of scientific measurements, such as the cosmic microwave background. The precision of these measurements needs to be very good (better than one part in 100,000) to get the amounts of dark matter and dark energy (within this model) to an uncertainty of better than 1%. The key point is that one can trace the measurement uncertainties back to the original observations and only then can one know whether to trust the scientific claim.

With that in mind, let's turn our attention to the matter-antimatter asymmetry of the universe. This is one of the most perplexing observations of Nature: there is more "matter" in the universe than **antimatter**. Before delving into the high precision data that supports this claimed asymmetry, let's start by describing what antimatter is and why we might expect symmetry (or balance) in Nature for matter and antimatter. Yet our mere existence depends on there being an imbalance of matter and antimatter!

3.1 What Is Antimatter?

According to the fundamental laws of particle physics, often called the Standard Model, matter and antimatter are treated on equal footing. When matter is produced, say from the energy of **photons** (particles of light), an equal amount of antimatter is also produced. Similarly, this process can be viewed in reverse, so that if matter particle meets up with an antimatter particle of the same type, the two annihilate, creating pure energy in the form of photons (see Figure 3.1).

The properties of antimatter are, according to Dirac's equation (see Chapter 2), a "mirror" of the matter properties. For example, the antiparticle of the electron, called the **positron**, has the opposite electric charge but is otherwise identical to an electron. In this case, the antimatter "mirror image" has reversed the electric charge (just like a mirror will reverse left and right—try holding up your left hand in front of a mirror and you'll see the image of yourself putting up its right hand). Both electron and positron have the same mass and radius. They are identical except for their

---

[1] See Chapter 1.

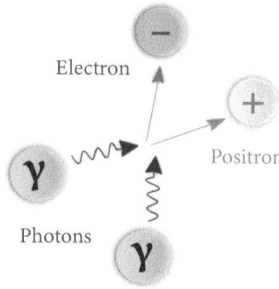

Figure 3.1 The creation of a matter-antimatter pair (in this case, an electron-positron pair) from photons.
Reproduced from Lobet (2015). Via Wikimedia Commons (CC BY-SA 4.0).

opposite charges. Similarly, an **antiproton** has a negative charge, opposite of the proton's positive charge.

If one takes an antiproton and a positron, it is possible to form **antihydrogen**. In regular hydrogen, the proton is at the center, with an electron orbiting it; in antihydrogen, the antiproton is at the center with a positron orbiting it. Does antihydrogen have the same properties as hydrogen? Since the physics (embodied by Dirac's equation) treats matter and antimatter the same, the internal properties of antihydrogen are the same. For example, the radius of the electron in hydrogen is the same as for the positron in antihydrogen. Since opposite charges attract each other, the two exhibit a kind of symmetry. Antihydrogen is the "mirror image" of hydrogen (see Figure 3.2).

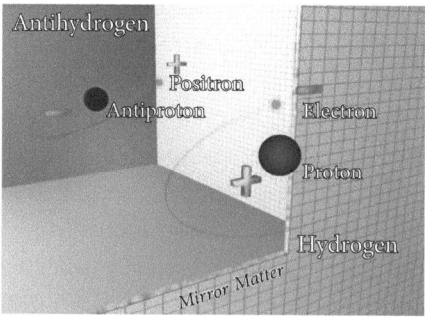

Figure 3.2 The hydrogen atom and its "mirror" image made of antimatter. In this fictional mirror, the sign of the charge is interchanged, too.
Reproduced from NSF (2011). Via Wikimedia Commons (public domain).

Antihydrogen is now routinely made using a source of antiprotons from the high-energy accelerator called **CERN** (Center European for Research Nuclear). Although the first antihydrogen atoms were produced in 1995, advances in technology were needed before the quantum energy levels in antihydrogen could be measured. In 2016, the transitions between quantum levels in antihydrogen were shown to be, within the measurement accuracy, identical to those in regular hydrogen (Ahmadi et al., 2016).

There is, of course, the difference in the charge between the electron and the positron. If you shoot an electron into a magnetic field, it will bend one way, whereas a positron will bend the opposite way. For example, if you shoot an electron into a magnetic field and it turns left, when viewed in a mirror the scene is reversed and it appears that the electron's image turned right. This is because mirrors swap left and right. Similarly, a particle bending to the left, when viewed in a mirror, appears to be bending to the right. So, if you observe the motion of an electron in a mirror, it will look identical to the motion of a positron. Note that you can't actually "see" the charge of a particle. There is no plus sign on the side of a positron! So, we deduce the charge of a particle by its interaction (bending left or right) with electric or magnetic fields.

There is a mathematical theorem that describes the relationship between matter and antimatter, called the **CPT theorem** (Lüders, 1954; Pauli, Rosenfelf, and Weisskopf, 1955), where the C stands for charge reversal, P stands for **parity** (a mathematical term for reversal of spatial directions—essentially interchanging left and right), and T stands for time reversal. The gist of the CPT theorem is easily understood: particles and antiparticles act like mirror images of each other.

So now we come to the main point. If Nature treats matter and antimatter as mirror images of each other, in the Big Bang there should have been equal amounts of matter and antimatter created. This leads to a universe filled half with matter and half with antimatter. The problem is that all observations so far indicate that the universe is dominated by matter. So, what happened to the antimatter from the Big Bang?

3.2 Where's the Missing Antimatter?

The evidence that our galaxy, the Milky Way, is made of matter (meaning that the stars, planets, dust, etc. are made from protons, neutrons, and electrons) is from samples of **cosmic rays**. The term "cosmic rays" was coined long ago, before it was known that these are just high-energy particles (mostly protons) that come from outer space. It is believed that cosmic rays originate from supernovae and other galactic sources (Ackermann et al., 2013). So, protons outnumber antiprotons by 4 orders of magnitude, or 10,000 to 1. The small numbers of antiprotons are probably the result of collisions of high-energy protons with galactic dust and not from a stellar antimatter source.

Additional evidence that the universe is dominated by matter comes from the fact that we don't see evidence of large-scale matter-antimatter annihilation (Figure 1 reversed in time). Consider what would happen if the nearest galaxy, Andromeda, were made of antimatter. In the mostly empty space between the galaxies, there is about one atom per cubic centimeter. This is a very small density, but the size of space is huge. It is easy to show that there would be lots of matter-antimatter annihilations, giving off telltale **gamma rays** (i.e., high-energy photons) that could easily be detected from Earth. Since we don't see these gamma rays, we can deduce that nearby galaxies are made of matter. Similarly, for galaxies farther away, the gamma rays would be less intense, but orbiting satellites with gamma detectors[2] would have seen this (Cohen, De Rujula, and Glashow, 1998). As a result, we conclude that the universe is made up of matter.

So again, why aren't there equal amounts of matter and antimatter from the Big Bang? It's not likely that the Big Bang produced only matter, as that scenario would violate the laws of physics. Instead, it's conjectured that the asymmetry we see today is the result of a tiny excess of matter over antimatter in the Big Bang, about one part in ten billion, and the rest of the antimatter annihilated within the first few milliseconds of the Big Bang (see Figure 3.3). This is an exquisite balance!

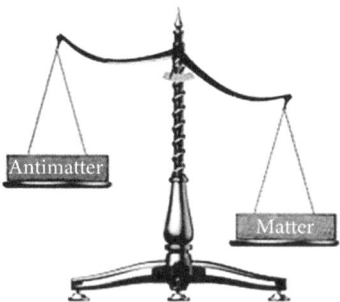

Figure 3.3 Balance scale showing the concept that matter in the universe outweighs the antimatter. At the time of the Big Bang, the excess of matter was tiny (an excess of only one part in ten billion).
Reproduced from www.fnal.gov/pub/today/archive.html Fermilab Today, October 31, 2013.

The delicate balance of equilibrium between matter and antimatter was somehow upset in the earliest moments of the Big Bang. How this happened

[2] For example, NASA's Fermi Gamma-ray Space Telescope.

is still a mystery. Although it would have been difficult to measure the excess of matter over antimatter during the Big Bang, the task is much easier now. Today, all the antimatter has been annihilated, and only the small excess matter is left. After this annihilation, the scale now has one part matter on the right and almost zero antimatter on the left.

Imagine if exactly equal amounts of matter and antimatter were present at the first moments of the Big Bang. In a short amount of time, the matter and antimatter would find each other and annihilate, leaving nothing but lots of photons. Of course, our world is made of matter, so human life could not arise without this tiny imbalance in the laws of physics for matter and antimatter. Clearly, it's important to understand what happened; after all, our very existence depends on this small effect.

Many experiments have been done at particle accelerators around the world, but so far the results cannot yet explain the matter excess of the universe. All measurements, even very precise ones (at an accuracy that exceed one part per trillion) show that matter and antimatter are produced in equal amounts. Next let's take a brief look at some of the experiments that are searching for an answer to this excess.

3.3 The Mystery of the Missing Antimatter

The laws of physics, as we currently understand them, are probably not complete. For example, we don't know whether dark matter is a particle (since no experiment has seen collisions between dark matter and regular matter). So, we don't include dark matter in the standard list of fundamental particles.[3]

A lot of new knowledge has been added to the field of physics over the past decades, but there is still a lot that we don't know. Currently, we don't know the root cause for the matter-antimatter imbalance, but an answer may be just on the horizon.

To distinguish between what is currently known and any new physics, the laws of particle physics are collected into a set of equations that are called the **Standard Model** (see Box 3.1), sometimes abbreviated as SM. The SM includes all known forces between particles: gravitational, electromagnetic, nuclear, and weak-nuclear forces. Virtually all experiments that have been done can be explained using the equations of the SM. Any new physics, such as an equation that would explain the excess of matter over

---

[3] See Chapter 4 for a listing of fundamental particles.

antimatter in the Big Bang, would be considered Beyond the Standard Model, or BSM for short.

Box 3.1 The Standard Model of Particles

When a mathematical model of Nature (often called a theory) is highly successful, having passed many tests from experimental measurements, it gets called a **Standard Model (SM)**. For particle physics, that framework includes a small number of fundamental particles (listed in Chapter 4) and a set of rules for how those particles interact with each other.

For example, there are six quarks in the SM of particle physics, along with the corresponding six antiquarks. All quarks (including antiquarks) interact by the exchange of gluons, providing the "glue" that holds particles (and atomic nuclei) together. The theory for quark-gluon interactions is called quantum chromodynamics. The "chromo" prefix references color, because the rules largely follow those of mixing the three primary colors of light (red, green, and blue). The naming of the particles and their interaction is a bit whimsical and serves as a mnemonic for those just learning about particle physics.

Part of the SM rules includes how particles and antiparticles are related. For example, if a particle has a property such as charge, the antiparticle has the opposite property. For example, the electron has a negative charge, so its antiparticle (the positron) has a positive charge. When a particle and its antiparticle come together, the two charges cancel, giving zero charge.

As another example of rules for antiparticles, quarks have a property that we call "color charge" (or just color). The quarks can't *actually* be seen with a color, but they have a *quantum property* that we simply label using colors. The antiparticles have the opposite colors. Bringing together a quark with a given color and an antiquark with its anti-color gives white (or colorless).[4] Just like the example above with charge, the two colors cancel, giving zero color.

There isn't room here to describe fully the SM, but the main point is that the subatomic world can be understood in terms of a handful of fundamental particles and their interactions. The SM may not be complete, but it seems to explain about 99.99% of experimental measurements. One thing it can't explain is why the Big Bang produced a slightly off-balance amount of matter and antimatter, resulting in an excess of matter that became our universe. The quest to understand that imbalance is one of the most important goals of present-day research.

Another facet that isn't included in the SM is dark matter and dark energy. No experimental measurement has shown evidence for dark matter, but we infer that it must exist due to astronomical observations (more on this later). If we ever find an explanation for dark matter, such as a new class of particles, it will likely be added to the SM. That quest is the "holy grail" of particle physics.

---

[4] Projecting two colors of light, such as red and cyan, together on a screen gives white light. Your TV screen creates a whole range of colors by utilizing the rules of color mixing.

The simplest form of the SM is one that has an exact (mirror) symmetry between matter and antimatter. The fact that the universe is dominated by matter implies there must be a new term in one of the SM equations that favors matter over antimatter. There have been many ideas of how this modification might come about, and even some ingenious experiments that have shown slight differences between matter and antimatter (such as measuring the weak decay of a particles called the neutral kaon or the B-meson (LHCb, 2017)), but this cannot yet explain the amount of matter-antimatter asymmetry in the Big Bang.

One of the most intriguing ideas to explain the matter-antimatter asymmetry involves a particle called the **neutrino**. This particle was first proposed by Wolfgang Pauli (see Chapter 2) in 1930 to explain why experiments looking at radioactive decay appeared to not conserve energy. At the time, conservation of energy was a standard part of all physics equations, essentially saying that "energy in" should be equal to "energy out." The conservation of energy is one of the reasons that a perpetual motion machine cannot be built; you can't get more energy out than you put in, and some energy is always lost to friction. So, conservation of energy was (and is) a fundamental rule in physics. Yet some energy seemed to be lost in the process of radioactive decay.

Pauli argued that the experiments were missing energy because an unknown particle, the neutrino, passed through their detectors without leaving a signal. At the time, Pauli's proposal seemed preposterous. Pauli even agreed, saying, "I have done a terrible thing. I have postulated a particle that cannot be detected." Of course, today we know that neutrinos exist (due to advances in particle detector technology). These ghostly particles can be detected with very large devices built exclusively for seeing the extremely rare collisions between neutrinos and matter.

A new and emerging idea is that the neutrino is its own antiparticle, meaning that the neutrino and the antineutrino are one and the same. This can only happen because the neutrino has a neutral charge. The reverse of a zero charge is still zero. This idea was first proposed by Ettore Majorana in 1937, and hence particles that are their own antiparticle are identified with his name. It is still controversial whether neutrinos are Majorana-type particles or not (see Box 3.2). At a more detailed level, it is also necessary that the "spin" of the neutrino (meaning its intrinsic angular momentum) obey certain conditions under the CPT theorem. Perhaps surprisingly, Majorana's neutrinos fulfill these conditions (Kayser, 1985).

Box 3.2 Majorana Particles

Ettore Majorana was reclusive but nonetheless a brilliant physicist. He worked with Enrico Fermi, Nobel laureate and legendary for his work on the Manhattan Project, who said, "There are several categories of scientists in the world; those of second or third rank do their best but never get very far. Then there is the first rank, those who make important discoveries, fundamental to scientific progress. But then there are the geniuses, like Galilei and Newton. Majorana was one of these" (Zichichi 2006). Fermi also remarked, at an executive committee during the Manhattan Project when they were at a critical point in that pursuit, "If only Ettore were here" (Zichichi 2006). Majorana had disappeared mysteriously from Italy in 1938, during events leading up to World War II.

Although brilliant, Majorana was not prolific. He published only nine papers and shunned the opportunity to take credit for discoveries he made. For example, when working in Fermi's group in 1932, he discovered the neutron based on experimental results from France (the French authors incorrectly interpreted the data) but never published according to Italian physicist Emilio Segré (Zichichi 2006). James Chadwick, an English experimental physicist, did a definitive experiment to show the neutron was real and was awarded a Nobel Prize for this.

Majorana's last paper, in 1937, was ground-breaking. He discovered an equation, like Dirac's equation, whose solution gives a *particle that is its own antiparticle*. In Dirac's case, the particle and the antiparticle are different, having opposite charges. In Majorana's solution, the particle can have neutral charge (such as a neutrino) and hence there is only one particle. To this day, there is ambiguity about the nature of the neutrino: is it a Dirac-like particle or is it a Majorana particle?

To solve this dilemma, which can only be answered by experiment, a worldwide effort is underway. Physicists may have found a path to the answer, but the problem is that it requires new technology (and hundreds of million dollars) to do the experiments, called neutrinoless double beta decay (see Box 3.3).

The neutrino may hold the key to how the excess matter resulted after the Big Bang. If the neutrino is a Majorana particle, then this could provide an imbalance in how matter and antimatter interact. It may explain why we now have a matter-dominated universe. However, the jury is still out. It will likely take a decade or more of experimental work before we get the verdict.

Regardless of how things turn out, Ettore Majorana has made brilliant contributions to theoretical physics. He formulated his equation long before anyone knew about the problem of matter-antimatter imbalance.

If the neutrino is a **Majorana particle**, then this could explain the matter-antimatter asymmetry of the Big Bang. This is one of the most popular hypotheses currently being discussed to explain the matter excess. One way to test this idea is to search for a rare type of radioactivity called **neutrinoless double beta decay**. That's a mouthful, but in essence this experiment looks for radioactive decays where no neutrinos are given off, in contrast to Pauli's idea that neutrinos are always released in radioactive decays. Then to conserve energy, all the energy must be used up by the emission of two electrons. (Historically, electrons were called beta particles, hence double beta decay.) If this very rare type of radioactivity is detected, it would prove that the neutrino is a Majorana particle. This would have far-reaching consequences, requiring a change to the SM and possibly explaining the matter-antimatter asymmetry.

To look for this new type of radioactivity requires very large detectors built specially for this purpose (see Box 3.3). The technology to do this is still being developed, and at present this is a worldwide effort involving scientists from many nations. It will likely take a decade of research before we will know whether the neutrino is a Majorana particle. It is one of several directions being pursued to explain the matter-antimatter imbalance.

Box 3.3 Double Beta-Decay Experiments

Regular beta decay is when a radioactive nucleus emits particles to become more stable. The nucleus radiates one electron plus one antineutrino (or, for some nuclei, their antiparticles). But in rare cases, the nucleus can emit two electrons plus two antineutrinos at once. This process, called **double beta decay**, can happen when the energy difference between the initial and final nuclei are favorable. So, double beta decay holds a unique place in radioactivity, allowing one to test ideas about neutrino properties.

In the SM, the neutrino and the antineutrino are regarded as separate particles. However, unlike the case of the other fermions in the SM (such as the electron), the neutrino has no charge. So, the neutrino particles are unique in this regard. How, then, do you distinguish between a neutrino and an antineutrino if they have the same (zero) charge? To answer this, we must delve a bit deeper into the SM.

continued

Box 3.3 *continued*

The equations of the SM assume that matter and antimatter are treated on equal footing. So, if a quark is produced (e.g., by high-energy collisions of particles), then an antiquark is produced at the same time. Similarly, if a lepton (this is a grouping that includes both electron-like particles and the neutrinos) is produced, then an anti-lepton must accompany it. For example, if an electron is emitted by radioactive decay, then an antineutrino must also be emitted.

In double beta decay, the SM rules say that two electrons are emitted, and so two antineutrinos must also be spewed out from the nucleus, and all four particles share the energy of the radioactive decay. However, if the neutrino is its own antiparticle, then one antineutrino can annihilate with the other, allowing two electrons to be emitted and **no neutrinos**. Here is the crux of the issue: if neutrinoless double beta decay is possible, then *the two electrons carry all of the energy*! Contrast this situation with regular beta decay, where some of the energy is carried away by the antineutrinos. These two cases can be clearly distinguished by experiments that measure the electron energies with high certainty.

The reason that so much effort is put into this topic is because it gives a possible explanation for the slight imbalance in the matter-antimatter from the Big Bang. The existence of our universe would not be possible without this imbalance. Currently, we don't know whether the matter excess of the universe is due to the neutrino being of the Majorana type (its own antiparticle), but we hope to have an answer soon from these ongoing experiments.

Although it's easy to say that measurements can distinguish between Dirac and Majorana neutrinos, the experiments are not easy. It has taken decades of work by some of the best experimental physicists in the world to develop the technology where these measurements are possible.

3.4 Antiparticles from Space

Another intriguing experiment looking at the matter-antimatter asymmetry is called the **Alpha Magnetic Spectrometer**, or AMS experiment, which is located on the International Space Station. Taking data since 2011, AMS measures the energy of particles that come from sources outside of our solar system. Most of these high-energy particles are protons, forming the bulk of cosmic rays, but a significant amount is also electrons and positrons. The AMS can measure the ratio of positrons to electrons (the positron fraction) and finds a much larger ratio at higher energies than predicted by calculations of known processes (shown by the curve in Figure 3.4). While AMS does not directly measure the source of the excess positrons, it does show that we don't fully understand how high-energy electrons and positrons are produced in outer space.

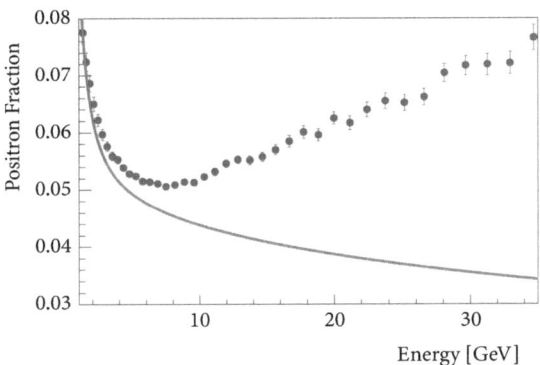

Figure 3.4 The faction of positrons, shown by the points, out of all electron-positron tracks measured at the Alpha Magnetic Spectrometer (AMS) versus the track's energy. The solid curve indicates the expected fraction from collisions of cosmic rays with residual interstellar gas at the orbit of the International Space Station, where the AMS is located.

Reproduced from AMS-02 Collaboration (2014). AMS-02 Press Release Sept. 18, 2014. ams02.space/news? field_news_taxonomy_target_id=64 (public domain)

The excess of positrons seen by AMS is baffling. One way to produce these particles is for a gamma-ray to interact with a proton, producing an electron and a positron in equal amounts. As the positrons move through space, many of them will annihilate with the electrons in hydrogen atoms that are found (with very low density) in space. But this process is part of the calculation that predicts a smooth decrease in the positron fraction above an energy of 10 GeV. Either we have a poor understanding of the environment in outer space or something else is going on that involves physics BSM.

Speculations abound for the source of the excess positron fraction seen by AMS. One exciting possibility is that it could be caused by annihilations of dark matter particles. Dark matter is thought to be made from particles that, like the neutrino above, have a very weak interaction with ordinary matter. Right now, there are trillions of neutrinos streaming through your body, and of course you don't feel it because these ghostly particles almost never interact with matter. The same goes for dark matter if current thinking is correct. While we don't know whether dark matter is made from yet-undiscovered particles, some theoretical physicists have speculated that dark matter particles could sometimes collide together, producing an electron and a positron in the process. Since the positron fraction is low to start with, a nearby source of positrons (from dark matter annihilation) could significantly change the positron-to-electron ratio at high energies.

Our ignorance of dark matter and how it might interact (with itself or with normal matter) is an indication of how much more scientists have yet to learn about the universe. Dark matter might be the source of the matter-antimatter asymmetry, but this is just speculation. In any case, AMS is clearly on to something, and their results are one of the new pieces of information recently that hint at new physics Beyond the Standard Model.

3.5 Why Does This Matter?

The mere fact that our existence depends on having more matter than antimatter in the Big Bang is one example of the delicate balance (or in this case, imbalance) in Nature. If the equations of the SM had their simplest form, with exact mirror symmetry for matter and antimatter, then we would not be here. But the reason for this imbalance still isn't understood by physicists.

The bottom line is that there is so much more to learn about the laws of physics. Future generations of scientists have a bright road ahead as they learn about the conditions that allow human life to exist. Even cartoonists have found the humor in where the antimatter has gone (see Figure 3.5).

Figure 3.5 Sidney Harris cartoon.
Reproduced with permission.

References

Ackermann, M. et al. (2013) "Detection of the characteristic pion-decay signature in supernova remnants," *Science*, 339, pp. 807–811.

Ahmadi, M. et al. (2016) "Observation of the 1S–2S transition in trapped antihydrogen," *Nature*, 541, pp. 506–510.

Cohen, A. G., De Rujula, A., and Glashow S. L. (1998) "A matter-antimatter universe?", *The Astrophysical Journal*, 495, pp. 539–549.

Kayser, B. (1985) "Majorana neutrinos," *Comments Nucl. Part. Phys.*, 14, pp. 69–86.

Lüders, G. (1954) "On the equivalence of invariance under time reversal and under particle-antiparticle conjugation for relativistic field theories," *Kongelige Danske Videnskabernes Selskab, Matematisk-Fysiske Meddelelser*, 28, pp. 1–17.

LHCb Collaboration (2017) "Measurement of matter—antimatter differences in beauty baryon decays," *Nature Physics*, 13, pp. 391–396.

Pauli W., Rosenfelf L., and Weisskopf V. (1955) *Niels Bohr and the development of physics*. New York: McGraw-Hill.

Zichichi, A. (2006) "Ettore Majorana: genius and mystery," *CERN Courier*, 24 July.

Chapter 4

Neutrons Matter

Young man, if I could remember the names of these particles, I would have been a botanist!

—Enrico Fermi, said to Leon Lederman

There are many things that we take for granted each day. The sun comes up in the morning, the stars come out at night, and the sky is blue (well, at least on Earth). Have you ever stopped to wonder what would happen if we could make a slight tweak to the laws of physics?

Let's do a thought experiment by making a slight change to Nature. We know that the nucleus is made up of protons and neutrons, with the mass of the protons nearly the same as the neutron. In our world, the proton is about 0.1% lighter than the neutron. In our thought experiment, let's imagine a universe where it's the other way around, with the neutron 0.1% lighter than the proton. Does it matter if one is slightly heavier than the other? As we'll see, the answer is a definitive yes!

In the real world, neutrons by themselves (not bound into a nucleus) will change into a proton with a half-life of about fifteen minutes. In other words, if you free a neutron from the nucleus, and wait on average for about fifteen minutes, the neutron will spontaneously change into a proton, emitting an electron (and another particle, an antineutrino) in the process (see Figure 4.1). The physical process is called neutron decay, and it's a type of radioactivity known as **beta-decay**. This process is also useful for applications, like in archeology (see Box 4.1). The bottom line is that a neutron can morph into a proton, but not the other way around. If you isolate a proton, it will stay that way for longer than the age of our universe (Nishino et al., 2009).

Nature's Balancing Act. Ken Hicks, Oxford University Press. © Oxford University Press (2025).
DOI: 10.1093/9780197771471.003.0004

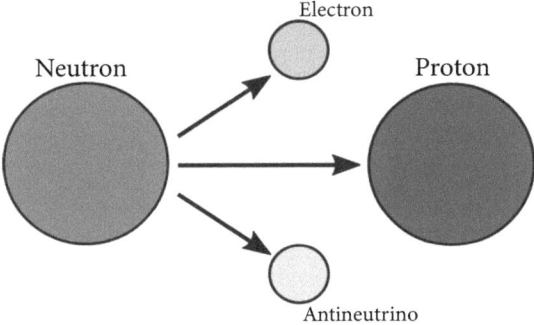

Figure 4.1 Radioactive decay of an isolated neutron, resulting in a three-particle final state of a proton, an electron, and an antineutrino.
Reproduced from Maxmath 12 (2021). Via Wikimedia Commons (Creative Commons CC0).

Box 4.1 Radiometric Dating

Radioactive nuclei decay into other nuclei with a characteristic lifetime, so the age of many materials can be determined from their natural radioactivity. This is useful in finding how long that material has been dormant, such as the age of fossils or the age of the Earth.

For example, radioactive carbon-14 is an isotope that occurs naturally, along with the more abundant carbon-12, which is nonradioactive (or stable). The difference is that carbon-14 has two more neutrons in its nucleus than carbon-12. When carbon-14 decays, it gives off an electron and an antineutrino, leaving a nucleus of stable nitrogen-14. When an organism is alive, it takes in carbon-14 from the air (or absorbs it from eating plants). After that organism dies, the carbon-14 slowly decays with a half-life of 5730 years. The result is that the ratio of carbon-14 to carbon-12 changes over time, allowing the date of the fossil to be calculated.

The carbon dating method is good for only a few tens of thousands of years. After that time, virtually all the carbon-14 has decayed away, and it becomes very difficult to detect. However, other radioactive isotopes have longer lifetime and can be used to date some types of minerals. For example, some rocks contain the element rubidium, and a naturally occurring isotope is rubidium-87, which decays to strontium-87, with a half-life of 50 billion years. By measuring the ratio of these two isotopes in rocks, the ages of the Earth and moon have been calculated.

The science of radiometric dating is complex, and many cross-checks with different types of radioactive decays must be done to accurately determine the age of a fossil or a rock. By now, scientists have well-studied methods that are generally

continued

Box 4.1 *continued*

---

accepted. These methods show that Earth is about 4.5 billion years old. In principle, anyone with the proper equipment can verify this.

The science of radiometric dating was first used in the early 1900s, soon after radioactivity was discovered. Before that time, many people thought that the age of the Earth was about tens of millions of years or less. Without radioactivity, people might still be arguing about the ages of fossils or moon rocks. Indeed, some biblical scholars in the past dated the Earth's age at about 6,000 years old. However, today's Catholic Church has no official position on this matter.

Radiometric dating is a wonderful example of how the interconnections between different branches of science (archeology and physics) lead to a dramatic increase in knowledge. Nuclear physics has touched our lives in different ways, whether it's the age of the Earth or radiation treatments for cancer.

In our imaginary world, the neutron would be lighter than the proton. In that case, the laws of physics stipulate that the proton would decay into a neutron (also emitting a positron and a neutrino). Well, if this actually happened, then after the Big Bang all free protons would convert to neutrons, with none left to form hydrogen. With no hydrogen, our sun would not shine (since hydrogen fusion fuels it). In this imaginary world, with just a *slight change* to the proton-neutron mass difference, it would be a cold, dark place.[1]

4.1 Quarks

What may have seemed like a trivial fact (that the proton is slightly lighter than the neutron) now takes on more importance. So, *why* is the proton lighter? It turns out that the proton is made up of three smaller particles called quarks, and quarks come in several varieties (also called *flavors*) and are given whimsical names, such as *up* and *down* (used only to label the quarks—with no directional significance to the names.) In the **quark model**, the proton is made from two *up* quarks and one *down* quark (see Figure 4.2). In comparison, the neutron is made from two *down* quarks and one *up* quark. You can think of the proton and neutron as kind of mirror images of each other, except our "mirror" now reflects an *up* quark into a *down* quark. Because the *up* quark is slightly lighter than the *down* quark, the proton is

---

[1] Some helium nuclei would form in this imaginary world, but helium does not undergo fusion easily (see Chapter 5). Helium stars could form and shine, but that would be a much different universe from the one we live in.

lighter. Of course, this only transfers the question, as now you could ask why the *up* quark is lighter. For that, there is presently no explanation (Hogan, 2000).

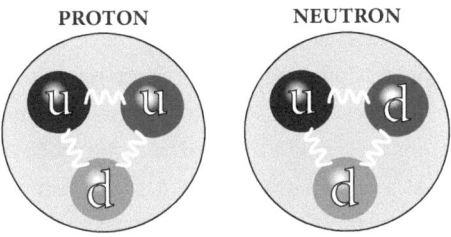

Figure 4.2 Quark substructure of the proton and neutron, in the simplistic quark model. The proton has two up quarks and one down quark, and vice versa for the neutron. The wavy lines connecting the quarks represent the strong force that binds them together.
Reproduced from Harp (2016). Via Wikimedia Commons (CC BY-SA 2.5).

In any case, our universe has *up* quarks that are lighter than *down* quarks. This has far-reaching consequences, because if it would be the other way around, then our universe would not form stars like our sun and could not harbor life as we know it. This delicate balance that tips the mass scale toward a heavier neutron appears even more coincidental when we look at the full spectrum of quarks and their masses.

Box 4.2 A Brief History of Quarks

The name quark was first proposed by Nobel laureate Murray Gell-Mann. He was one of the first to propose that the proton had a substructure. He got the name from a novel by James Joyce where the phrase "three quarks for Muster Mark" appears. Gell-Mann, who has a sense of humor, liked this nonsense word and presumably chose it because his model of the proton had three subunits, which he whimsically called quarks. In his book *The Quark and the Jaguar*, Gell-Mann writes, "The recipe for making a neutron or proton out of quarks is, roughly speaking, 'take three quarks.'"

The mathematical model that Gell-Mann used (based on group theory) was his way of organizing a plethora of new particles that had been discovered at particle accelerators in the 1960s. These particles typically had masses that were slightly heavier than the proton, and some particles had a neutral charge (like the neutron, but heavier) or negative charge. It's not clear whether Gell-Mann believed that quarks were real particles, or whether his model was just a convenient mathematical tool, but

continued

Box 4.2 *continued*

---

later experiments done at the Stanford Linear Accelerator (SLAC) showed that quarks are indeed real.

The SLAC experiments fired high-energy electrons at protons and looked at how those electrons scattered (called deep inelastic scattering). If the proton was simply a diffuse ball of charge, the electrons should be slightly deflected. However, the experiment measured electrons bouncing backward from the proton. To help visualize this, think of firing bullets at a snowbank. If it's just snow, then the bullet will go right through. If it ricochets back, though, then it must have hit something small and hard, like a stone inside the snowbank. The quarks are analogous to the stone, and electrons in place of bullets. The clear result of the SLAC experiment was that there was something small and hard inside the proton, which they called "partons." Further comparison between the SLAC data and theory predictions showed that the partons were consistent with the quarks in Gell-Mann's model.

The quark model did more than predict that the proton had substructure. This model also predicted that there should be other particles like those just discovered, but heavier. One of these, now called the omega-minus, has roughly twice the mass of the proton and has a negative charge. It was predicted by the quark model in 1962 and first seen in experiments in 1964. Over the years, the quark model has been very useful in helping to categorize new particles seen from accelerator experiments, but today it is considered as too simplistic to describe the proton's structure (see Box 4.3).

Like many good ideas, the quark model was formulated independently, in this case by George Zweig, with important contributions by Yuval Ne'eman. Neither received the same level of publicity as Gell-Mann, but history shows that they deserve substantial credit for their contributions to the development of the quark model.

Quarks were proposed (independently) by two theoretical physicists: Murray Gell-Mann and George Zweig (Gell-Mann 1964, Zweig 1964) (see Box 4.2). In their original model, they had just three types of quarks: *up*, *down*, and *strange* (abbreviated as u, d, and s). The s-quark was needed to explain the existence of newly discovered particles that were produced in high-energy collisions of protons on protons in the 1960s, which had unusual decay properties. These particles were called, for lack of a better term, strange particles. When accelerators became more powerful, with higher energies, a new class of particles was discovered, whimsically called *charmed* particles. This called for another quark, the c-quark. With the addition of the c-quark, the list of quarks started to look more orderly. Both d and s quarks have an electric charge equal to one-third of the electron's charge, or −1/3, whereas

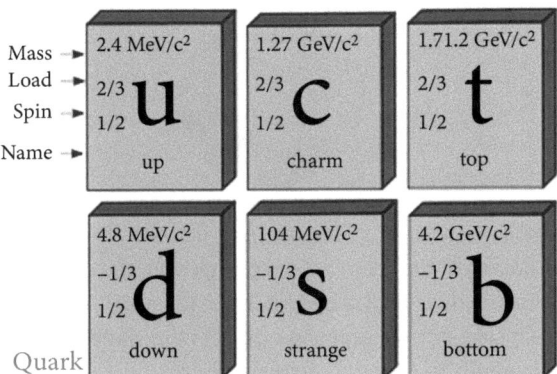

Figure 4.3 Table of known quarks, showing the quark symbol along with the mass and charge (or load), where the top row has charge +2/3 and the bottom row has −1/3.
Reproduced from Ranjithsiji (2019). Via Wikimedia Commons (CC BY-SA 4.0).

both u and c quarks have a charge of +2/3. After another decade or two, with further advancements in accelerator technology, two more quarks were found, one in each charge category, called *top* and *bottom* (or t and b quarks). No other quarks have been found (Workman et al., 2022), so we are left with six types of quarks: three with charge +2/3 and three with charge −1/3 (see Figure 4.3).

What is more interesting is the trend for the quark masses. Even though we don't yet know how to calculate (from first principles) the mass of any quark, that doesn't stop us from looking for some trends of the masses. Obviously, the weight of these tiny particles is far smaller than our typical units of weight, and it doesn't make sense to use units like pounds or kilograms. For convenience, scientists use a unit that is closer to the mass of these particles, called[2] the MeV and the GeV (where 1 GeV = 1,000 MeV). The numerical values of the masses are shown in Figure 4.3. Systematically, the quarks get more massive and can be grouped into three "families": the u and d quarks are lightest, with masses of a few MeV; the s and c quarks have masses range from about 0.1 to 1.25 GeV; and the b and t quarks are the heaviest family, with the t quark at a whopping 175 GeV (see Figure 4.4).

Note that for the two heavier families (the c-s quark family and the t-b quark family), the top row in Figure 4.3 has a much larger mass than the bottom row. The c quark is over ten times as massive as the s quark, and the t-quark is about forty times the mass of the b quark. Based on this trend, you might guess that the u quark should be more massive than the d quark. **But that's not the case!** The u quark is less than half the mass of the d quark. Why Nature should be

---

[2] MeV is short for Mega-electron-Volt, and GeV for Giga-eV, which technically are units of energy, but of course Einstein tells us mass and energy are interchangeable

Figure 4.4 Schematic drawing of quarks as spherical balls having the same density, to show the relative masses of the quarks. For comparison, the masses of the proton and electron (using the same density) are shown in the lower-left corner. In fact, the quarks do not have a known size, but are point-like objects too small to measure. This fictional drawing is simply to give a feel for the relative masses.
Reproduced from Mrsi (2012). Via Wikimedia Commons (CC BY-SA 3.0).

this way is a deep mystery and, as mentioned above, tipping the balance of masses this way keeps the neutron heavier than the proton (and hence our sun shining).

You may have noticed that adding the masses of the quarks making up the proton (two u quarks and one d quark) and comparing this with the sum for the neutron (two d quarks and one u quark), these two sums are more than 0.1% different. The reason for this is that the quarks are bound together in the proton (or neutron) and this binding energy is large. The binding energy adds to the mass (again, Einstein tells us that energy and mass are equivalent). In fact, about 99% of the proton's mass comes from its binding energy (Wilczek, 2003) (see Box 4.3). It turns out that the binding energy is roughly the same for the neutron and the proton, and so the neutron, with two d quarks, comes out slightly heavier (Walker-Loud, 2018).

Box 4.3 What Is Mass?

It sounds like a simple question: what is the mass of the proton? We can measure this in particle detectors (from its curvature as a proton as moves through a magnetic field), giving a mass of about 940 MeV. But if you add up the mass of the proton's three quarks, it's less than 10 MeV. So, what are we missing? The missing piece is binding energy, in this case from the nuclear potential energy. Recall $E=mc^2$, so energy is basically the same as mass. The energy that binds the proton together must be added to the quark masses. This means that about 99% of the proton's mass is binding energy.

If you weigh yourself, most of your weight comes from the nucleus inside the atoms in your body. Since the nucleus is made from protons and neutrons, you could say that about 99% of your "weight" is from energy, not mass! Here, I have assumed that the quarks and electrons have actual mass and don't have any substructure (as in the Standard Model).

When we use Newton's relation that force is equal to mass times acceleration (symbolized by the equation F=MA), the mass is what we measure on a bathroom scale. So, what is mass? It's not so simple. As Einstein taught us, mass is the same as energy, but since most of the luminous mass of the universe is from protons, 99% of that mass is energy. To understand where all the mass of the universe comes from, the biggest part of that tally is keeping track of the energy.

If the theory of inflation (see Chapter 1) is correct, then the mass energy is balanced by the (negative) gravitational energy. It requires a different way of thinking about our universe if most of the "mass" is just energy. Yet that is what we have learned in the past few decades about the proton's mass. Science is constantly moving forward, and our view of the proton's mass has changed drastically since the 1990s.

It is said that Nature likes to repeat itself. For example, we find many of the same features in the nucleus that we see in atoms, such as electrons organized into a "shell structure" with inner orbitals and outer orbitals, and similarly protons and neutrons form shell structures in a nucleus. Inside the proton, the quarks also organize into shell structures. The reason for this is that the laws of quantum mechanics apply to both bound systems.

What other examples are there of Nature repeating itself? We know that atoms are mostly empty space, meaning that the size of both the nucleus and the electron are miniscule compared with the size of the atom. If you could get a microscope powerful enough to see inside solid material, the spacing between atomic nuclei is about 100,000 times larger than the size of the nucleus, and the space between contains some electrons, but is otherwise empty.

Similarly, inside the proton, the spacing between the valence quarks is vast compared with the size of those quarks. One way of looking at the proton is that it, too, is mostly "empty" space! Here, the space is filled with a swarm of gluons that are exchanged between quarks, except that gluons have no mass. The old view of the proton as a solid "ball of mass" is simply not correct.

4.2 Leptons

In addition to the quarks, another category in the fundamental building blocks of matter has particles called **leptons** (see Figure 4.5). The name literally means "light particle," and this category includes the common electron, which is about 2,000 times lighter than the proton. The neutrino is also a lepton, which has such a small mass that it is extremely difficult to measure.

It turns out that there are three types (or families) of leptons, coming in pairs just like quarks. It is still unknown why there should be just three families of each pair of leptons. While scientists have searched for a fourth family, there is strong evidence (Decamp et al., 1989) that the progression stops at three.

Why are the leptons important? Because they round out the table of fundamental particles (see Figure 4.5) showing a common trend between quarks and leptons. This suggests that there is order to Nature, like that found in the periodic table used for chemistry. The periodic table provides information on the common chemical properties of atoms, with similar types of atoms arranged in columns. Similarly, the table of fundamental particles groups common aspects between quarks and leptons, hinting at a deeper connection between these two particle groups (Fritzsch, 1989).

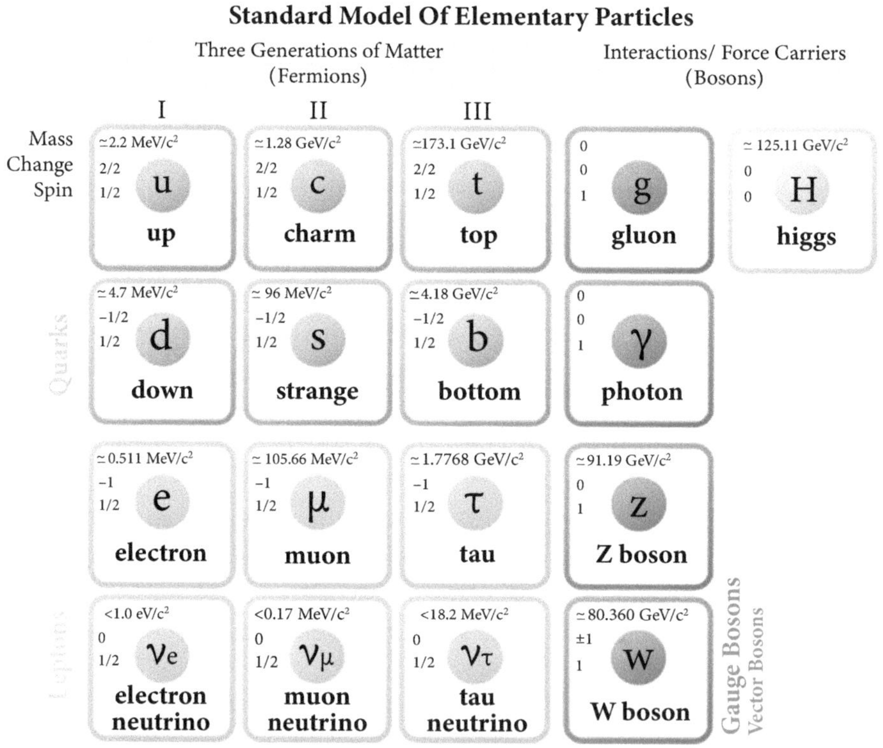

Standard Model Of Elementary Particles

Figure 4.5 Table of all elementary particles in the Standard Model, showing the three families of quarks and leptons (labeled I, II, III) and the boson particles associated with three of the known forces. The Higgs boson is also shown.
Reproduced from Cush (2019). Via Wikimedia Commons (public domain).

Before going further, let's examine the leptons in more detail. The first lepton pair is the electron and the electron-neutrino, the latter being a particle that is produced in radioactive nuclear decay (in conjunction with an electron being captured by the nucleus). The notation used for neutrinos is the Greek letter nu, written as ν, with a subscript showing its pair-produced partner, such as ν_e for the electron-neutrino. The second lepton pair has a particle called a muon (see Box 4.4), denoted by the Greek letter mu (μ) along with the corresponding muon-neutrino. Finally, the third lepton pair is given the Greek letter tau, or τ, along with its neutrino. There are various ways to produce neutrinos in the laboratory; one method is from collisions between high-energy particles (e.g., at particle accelerators). Combining leptons, quarks, and the particles responsible for the forces between particles (see the last column of Figure 4.5), we get the **Standard Model** of elementary particles.

Box 4.4 Discovery of the Muon

The discovery of the muon, a particle much like the electron, in 1936 was quite a surprise. At the time, the only known subatomic particles were the proton, electron, and neutron. The latter was only discovered a few years earlier but had been predicted from studies of nuclei. In contrast, Nature gave no clues that there could be a particle that was like the electron but much heavier.

To put this in perspective, we need to think like a scientist of the early 1930s. At the time, particle accelerators were not powerful enough to produce new particles from collisions of nuclei. Even radioactivity was a fairly new phenomenon. Normal matter is made up of atoms, having nuclei and electrons, so there was no reason for scientist to speculate that there were other particles.

Studies of cosmic rays (high-energy particles—mostly protons—that constantly bombard Earth from space) first showed that a particle existed with a negative charge and a mass 200 times heavier than the electron. Initially, this new particle was named the mesotron[3] by Carl Anderson, who first saw it in cosmic rays (with Seth Neddermeyer). Later, it was renamed the muon, following a naming convention constructed to categorize new particles.

Curiously, just a year before the muon was discovered, theoretical physicist Hideki Yukawa had predicted a new particle that could explain the strong-nuclear force. This particle, now called the pion, was just a hypothesis at first, as Yukawa had a novel way of solving Schrödinger's equation. So, when the muon was discovered, many scientists thought it could be Yukawa's particle. It even had a mass in the range predicted by Yukawa's model.

Further experiments showed that the muon didn't interact strongly with the nucleus and so couldn't possibly be Yukawa's particle. It seemed so coincidental that there was another particle in Nature with a similar mass that physicist I. I. Rabi was

continued

Box 4.4 *continued*

---

quoted to say, "Who ordered that?" Needless to say, no one could figure out a reason why the muon was deemed necessary by Nature.

Today, we see that the muon provides a symmetry in Nature between quarks and leptons, as shown in Figure 4.5, where the second column has a new "family" of particles that mimics the organization of the first column. Again, this hints at a deep connection between quarks and leptons that is still unexplained by current theoretical models.

---

[3] Meso is a Greek prefix meaning "middle," so named since its mass was midway between the electron's mass and the proton's mass.

Entire books have been written to describe the details of how these fundamental particles interact with each other. Here, the important point is that there seems to be a kind of ordering to the fundamental particles. For example, the neutrinos have zero charge, and the electron-like particles have one unit of negative charge, giving one unit of charge when going from the bottom row of leptons to the next higher row. Similarly, going from the lower row of quarks to the upper row, the charge differs by exactly one unit. This suggests a deep connection between the quarks and the leptons, a correspondence that is still not understood (even if there are some theoretical models—speculative at present—that purport to provide such a connection).

Once again, Nature has been kind. We take it for granted that the charge of the electron is exactly opposite to the charge of the proton. At present, we don't know why this should be, and one could imagine a world where the electron charge is some fraction of the proton's charge. Again, this would result in a very different universe, one where hydrogen (made from one proton in the nucleus and one orbiting electron) does not have a net neutral charge. In that case, stars and planets would not form due to the overwhelming electric force between atoms.

While it's likely that there is a deeper level of physical laws that connect the quarks and the leptons, such as the speculative string theory (see Chapter 2), at present we are just left to ponder about clues that Nature has provided to us. If there is more to learn beyond the Standard Model of particle physics, it may be that theoretical physics will lead the way, just as it did when Einstein gave his explanation of gravity as a geometric curvature of space-time. Maybe some mathematical equation can be found, making it inevitable that quarks and leptons are connected in some way. However, if history is our guide, it is more likely that experimental physics will find new elementary particles, which would lead to a revision of the Standard Model. This might happen at the world's largest accelerator facility (the **Large Hadron Collider**, see Box 4.5). These new particles, if they exist, may point the way to the interrelationship between quarks and leptons (Aitchison 2005). Only time will tell whether such particles can be found.

Box 4.5 The Large Hadron Collider

Over time, accelerators have gotten bigger and more powerful, with the consequence that it costs more and more to build them. The Large Hadron Collider (LHC) is the biggest accelerator on Earth, with a circumference of seventeen miles, and has particle detectors as big as multistory buildings. This was made possible due to monetary contributions from many countries around the world. It is located on the border between Switzerland and France, with headquarters in Geneva.

The main goal of the LHC is to collide protons at the highest energies possible, in this case about 10,000 GeV (where the proton mass is on the order of 1 GeV), with the hope of detecting new particles such as the Higgs boson and possibly more massive particles predicted by speculative theories such as supersymmetry. To date, the Higgs boson was discovered at the LHC, but no other new particles. The measured properties of the Higgs boson at the LHC have confirmed its prediction by the Standard Model of particle physics.

Since new particles predicted by the supersymmetry model have not been seen at the LHC, this leaves several open questions about the Standard Model. For example, is there a connection between quarks and leptons? Supersymmetry had an answer for this, but if supersymmetry isn't correct, then what is the right theory? Of course, it could be that the supersymmetric particles are more massive than originally thought. However, to see them would require an even bigger accelerator than the LHC.

In addition to proton-proton collisions, the LHC can accelerate the nuclei of heavy elements, such as lead (chemical symbol Pb). Smashing together two lead nuclei (called Pb-Pb collisions) has led to new discoveries about quarks and gluons. When the collision energies get high enough, quarks are no longer bound inside protons and create a "plasma" where quarks and gluons can, for a short period, roam freely like air molecules moving about the room. The "room" containing this plasma is the size of an atomic nucleus, but this is still large compared to the size of quarks. The quark-gluon plasma is a new state of matter, which exists only temporarily at these

continued

Box 4.5 *continued*

ultra-high temperatures, but stays around long enough that physicists can study it using the particle detectors at the LHC. These studies provide new information about the strong-nuclear force.

The LHC was built in the early 2000s and is nearing the end of its useful scientific life span. Already, plans have been made for an even bigger accelerator (with a larger circumference) to be built at the same location. Significant government funding from many nations will be needed for its success.

4.3 The Nuclear Force

Let's return to the Big Bang. In the first few seconds, all matter in the universe was a dense gas made of protons, neutrons, and electrons. As mentioned in Chapter 2, some of the neutrons were attracted (by the strong-nuclear force) to protons, forming a nucleus called a deuteron (labeled D, see the top of Figure 4.6). But many of the neutrons decayed into protons in the minutes following the Big Bang, leaving an excess of protons. As the universe cooled, those excess protons combined with electrons to form electrically neutral hydrogen atoms. Electrons also combined with deuterons to make neutral deuterium atoms. The primary difference between hydrogen and deuterium is that the latter is heavier—almost twice as heavy since deuterium has both a proton and a neutron in its nucleus. These atoms, over time, came together to form gas clouds that gravitationally collapsed into stars. Astronomers can measure the ratio of hydrogen to deuterium in very old stars.[4] The ratio of hydrogen to deuterium can also be calculated by using our current knowledge of nuclear physics coupled with the density of matter predicted by the Big Bang (Peebles, 1966). The measured ratio is in good agreement with that calculated by this theoretical model (Fields, Molaro, and Sarkar, 2021). This is strong evidence in favor of the Big Bang.

Other nuclei, in addition to deuterons, were also formed from the Big Bang. The amounts of these other nuclei can be calculated using the same theoretical model and compared with astronomical observations, providing further evidence for the Big Bang. For example, two deuterium can collide, forming tritium plus an excess neutron (Figure 4.6, third row), which can then

---

[4] Astronomers use the technique of spectroscopy to measure the ratios of atoms in stars. This technique uses the color spectrum given off by stars, which contains unique information about the elements in the star's body. Breaking starlight into a color spectrum requires large telescopes with impressive technology.

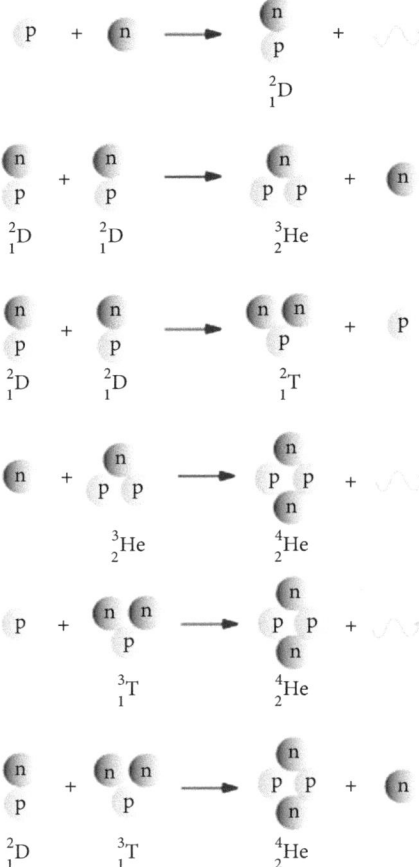

Figure 4.6 Schematic of nuclear reactions that took place during the Big Bang. The symbol D, T, and He stand for deuteron, triton, and helium, respectively. The superscript labels the total number of protons and neutrons in the nucleus, whereas the subscript counts only the protons (equal to the overall charge). The wavy lines indicate that energy is given off as photons.

Reproduced from Kosmider (2011). Via Wikimedia Commons (CC BY-SA 3.0).

form helium by absorbing a free proton (Figure 4.6, fifth row). Alternatively, deuterons can combine with helium's nucleus to form lithium, with three neutrons and three protons. A lot of nuclear physics goes into these calculations, with the result that astronomical observations of light nuclei (hydrogen, deuterium, helium, and lithium) match closely with the theoretical models of Big Bang nucleosynthesis (Fields, Molaro, and Sarkar, 2021). It is remarkable that nuclear reactions measured on Earth can predict the ratios of light elements we see in the stars!

There are subtle aspects of the deuteron that are also important. As mentioned above, the nuclear force provides the attraction between the proton and the neutron in deuterium. This is the reason that the protons don't just bounce off the neutrons like two billiard balls striking each other. Imagine two billiard balls, each with a bar magnet embedded in its center. If the balls strike each other slowly, the magnets will line up and be attracted to each other, causing the balls to stick together. Making deuterium from the collisions of protons and neutrons is something like this, except with the magnetic force replaced by the nuclear force.

It turns out that the proton and neutron also have the equivalent of a bar magnet inside, a property called spin (see Chapter 2). The magnetic force between the proton and the neutron is much smaller than the nuclear force, but still important. It turns out that the neutron and proton will become bound (i.e., stick together) only if their spins line up (Cappeliaro, 2022). This is somewhat analogous[5] to two magnets. If two bar magnets are lined up correctly, they will attract. Turn one of them around by 180 degrees, though, and they repel. Similarly, if the two spins anti-align, they repel, overpowering the nuclear attraction. In that case, the deuteron is no longer **bound**, allowing the proton and neutron to separate.

The amazing thing here is that while the proton-neutron system can bind together, a pair of two protons (or a pair of two neutrons) will not bind. This is because the Pauli exclusion principle (see Chapter 2) prevents the proton pair (or neutron pair) from having their spins aligned.

Now consider a universe where the nuclear force is just a little bit stronger. In this imaginary world, the force between the spins cannot overcome the attractive nuclear force and so two protons (or two neutrons) will bind together. This again would be a very different universe from the one we live in. In that case, after the Big Bang, neutrons and protons would combine more easily, drastically increasing the helium-to-hydrogen ratio (Barrow, 1987). Also, nuclei in stars would fuse together more quickly, reducing the lifetime of a star (Adams, Howe, Grohs, and Fuller, 2021). Some authors (Barnes, 2015) have suggested that life could not form in this imaginary universe where diprotons and dineutrons exist. This is not clear, though, because smaller-mass stars could now shine (Adams, Howe, Grohs, and Fuller, 2021). What is clear is that just a small increase (about 10%) in the nuclear force would change the landscape of our universe.

---

[5] The analogy is not perfect because the interaction of two spins is governed by quantum mechanics. So, any comparison with bar magnets is not the same. The idea here is that the orientation of the spins can change the force that holds the deuteron together.

Similarly, if the nuclear force was a bit weaker, the proton-neutron system could not bind into deuterium (Golowich, 2008). If deuterium was unbound, then the standard fusion process that provides sunlight would not happen. At first glance, this other universe (with a slightly weaker nuclear force) would appear to be a very cold and dark place, inhospitable to life. Indeed, some authors have suggested this. However, it was recently shown (Adams and Grohs, 2017) that a different fusion process could occur, called the **triple-nucleon** reaction, powering stars in this imaginary world. Whether life could exist in such a place is an open question. But still, just a small *decrease* in the nuclear force would result in a universe unlike our own.

In summary, the nuclear force appears to be **fine-tuned** (or balanced) in just the right way: to be strong enough that deuterium can bind, yet not strong enough that two protons can bind together. Change the strength of the nuclear force by about 10% either way and that universe would be unrecognizable.

You might ask why the nuclear force (or the proton-neutron mass difference) has the value that we find in Nature. Although this is a fascinating question, it is not one that science can yet answer. The relative strength of the forces in Nature—gravitation, electromagnetic, weak and strong forces— are values we get from measurement, not from any theoretical model. Perhaps someday an ultimate "theory of everything" will be discovered, but at present we must just accept that these constants of Nature are what they are. This is *another cosmic coincidence* that is necessary for our universe to be hospitable for human life.

The strength of the nuclear force is also important to produce heavier elements, such as carbon, oxygen, and iron. In Chapter 5, we'll find that the nuclear force, again, has just the right value to allow these elements—those necessary for life—to be produced in stars. As we go along this journey, exploring our solar system and beyond, we will find more and more cosmic coincidences.

References

Adams, F. C. and Grohs, E. (2017) "On the habitability of universes without stable deuterium," *Astroparticle Physics*, 91, pp. 90–104.

Adams, F. C., Howe, A. R., Grohs, E., and Fuller, G. M. (2021) "On the habitability of universes without stable deuterium," *Astroparticle Physics* 130, pp. 102584.

Aitchison, I. J. R. (2005) "Supersymmetry and the MSSM: An elementary introduction." Available at: https://arxiv.org/abs/hep-ph/0505105.

Barnes, L. A. (2015) "Binding the diproton in stars: Anthropic limits on the strength of gravity." Available at: https://arxiv.org/abs/1512.06090.

Barrow, J. D. (1987) *Phys. Rev. D*, 35, pp. 1805–1810.

Cappeliaro, P. (2022) "The deuteron." Available at: https://phys.libretexts.org/@go/page/25718.

Decamp, D. et al. (ALEPH Collaboration) (1989) *Phys. Lett. B*, 231, pp. 519–529.

Fields, B. D., Molaro, P., and Sarkar, S. (2021) Prog. Theor. Exp. Phys. 2022, section 24, pp. 1–15 and updates at http://pdg.lbl.gov.

Fritszch, H. (1989) *Quarks*. New York: Basic Books.

Gell-Mann, M. (1964) "A schematic model of baryons and mesons", *Phys. Lett.*, 8, pp. 214–215.

Golowich, E. (2008) "Unbinding the deuteron." Available at: arXiv:0803.3329.

Hogan, C. J. (2000) "Why the universe is just so," *Rev. Mod. Phys.*, 72, pp. 1149.

Nishino, H. et al. (Super-Kamiokande Collaboration) (2009) *Phys. Rev. Lett.*, 102, 141801.

Peebles P. J. E. (1966) "Primeval Helium Abundance and the Primeval Fireball," *Phys. Rev. Lett.*, 16, pp. 410–413.

Walker-Loud, A. (2018) "Dissecting the mass of the proton", *Physics*, 11, p. 118.

Wilczek, F. (2003) "The origin of mass," *MIT Physics Annual 2003*, pp. 24–35.

Workman, R. L. et al. (Particle Data Group)(2022) *Prog. Theor. Exp. Phys.*, pp. 083C01 (and updates at http://pdg.lbl.gov).

Zweig, G. (1964) *CERN Reports No. 8182/TH.401 and No. 8419/TH.412.*

Chapter 5
So Much Carbon

There are puzzles and problems in each part of the cycle which challenge the basic ideas underlying nucleosynthesis in stars.

—Willy Fowler, Nobel Prize Lecture, 1983

Imagine that you could play the role of "chef of the elements." Your goal is to create a recipe for a universe that would be hospitable to carbon-based life forms (as we have on Earth). You will need a variety of elements, more than just hydrogen and helium, which were created in the first few minutes of the Big Bang.

Apart from hydrogen, the human body consists of about 65% oxygen (by weight), 18.5% carbon, 9.5% hydrogen, and small percentages of many heavier elements. Those heavy elements, including those in the DNA molecule such as phosphorus, are considered necessary for life (Zoroddu et al, 2019). How were the heavier elements, like carbon and oxygen, made from the lighter hydrogen and helium? Before going over the recipe for making carbon, let's review some basics. The first element on the periodic table is hydrogen, with chemical symbol H. It's the simplest element, with one proton in the nucleus and one electron orbiting the nucleus. Recall that two protons don't bind together, but neutrons can bind with protons, forming deuterium, with chemical symbol ^{2}H. As shown in Chapter 4, deuterium can absorb neutrons and protons during the Big Bang to form helium (^{4}He), where the number in the superscript gives the total number of protons plus neutrons. But essentially no carbon (^{12}C) was produced in the Big Bang (Turner, 2022).

You might think it would be easy to continue this process of making heavier nuclei, called **nucleosynthesis**. But it turns out that the nuclear physics is not that simple. Nature gets stuck during the Big Bang. Temperatures cooled quickly after the Big Bang, and the possible nuclear reactions are limited. To see why this is, let's look at the first step in the fusion of light nuclei, shown in Figure 5.1, combining ^{2}H and ^{3}H.

Note that these nuclei don't stick together. The intermediate nucleus is unstable, lasting only a brief instant before breaking up into ^{4}He plus a neutron. In contrast, if a neutron collides with a proton (at a low energy), it sticks

Nature's Balancing Act. Ken Hicks, Oxford University Press. © Oxford University Press (2025).
DOI: 10.1093/9780197771471.003.0005

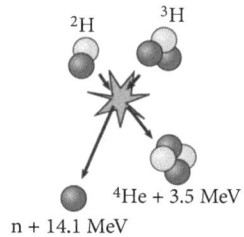

Figure 5.1 The fusion of deuterium and tritium, which collide together but cannot stick. The result is a helium nucleus (^{4}He) plus a neutron, each moving away with some kinetic energy (shown in units of MeV).
Reproduced from Wykis (2007). Via Wikimedia Commons (public domain).

to form ^{2}H. Adding another neutron gives the radioactive nucleus tritium (^{3}H), but adding one more neutron makes the system unstable, mainly due to the Pauli exclusion principle.[1] Similarly, adding ^{2}H to ^{3}H in Figure 5.1 gives a nucleus that isn't stable. The point here is that no nucleus with total of five **nucleons** (i.e., adding protons and neutrons) sticks around (Lauritsen and Ajzenberg-Selove, 1966). That spoils the chain of making heavier nuclei, which constrains the number of elements that can be made during the Big Bang.

As a chef of the elements, making a recipe for nucleosynthesis would be easier if a nucleus with five nucleons were possible. In that case, more elements could be formed during the Big Bang. A bound state of five nucleons could be made, in an imaginary universe, by tweaking the nuclear force a bit. But nature doesn't always take the easy path, and certainly didn't here. Similarly, there is no bound nucleus with eight nucleons, creating another bottleneck to nucleosynthesis. This makes the recipe for making heavier nuclei more difficult (but not impossible). Physicist George Gamow (see Box 5.1) and his collaborators first considered this problem back in 1950 (Alpher, 1948).

Sidebar 5.1 George Gamow

Born and educated in Russia, George Gamow and his wife defected from the Soviet Union due to oppression, such as being denied repeatedly to attend international

---

[1] Neutrons and protons have spin ½, and so no two neutrons can have the same quantum numbers. Two neutrons can have opposite spin direction, but a third neutron can't occupy the same nuclear orbit as the other two. The result is that the binding energy is decreased, making ^{4}H unbound.

conferences. In 1933, he was allowed to attend one of the famous Solvay Conference Series held in Brussels and eventually made his way to the United States. There, Gamow and his students did seminal work on the synthesis of elements, first published in 1948 in a landmark paper (Alpher, Bethe, and Gamow, 1948) that is often called the "alphabet paper" because the authors' names sound similar to the first three letters of the Greek alphabet: alpha, beta, and gamma. That paper was the first to calculate, from basic nuclear physics, the proportions of hydrogen, helium, and some heavier elements that would be created in the early universe following the Big Bang. Unfortunately, the calculation of heavier elements did not take into account that there is no stable nucleus with a mass of 5 (having a total of protons and neutrons that sums to five) or a mass of 8. Later, other physicists such as Fred Hoyle found that the heavier elements can be made in stars, via a process known as stellar nucleosynthesis.

Before leaving the Soviet Union, Gamow made fundamental contributions to the understanding of radioactive decay of heavy elements like thorium and radium, which naturally spew out alpha particles, one of three ways that radioactive nuclei can release stored energy (or "decay"). In 1928, quantum mechanics was relatively new, and Gamow applied one of the new ideas called quantum tunneling to radioactive decay. There is no analog in classical physics to quantum tunneling, which says that a particle can penetrate through a seemingly impossible barrier. Yet this happens and is the basis for many subatomic phenomena, and this principle is at the heart of quantum computing. The use of quantum tunneling to explain features of radioactive decay, such as the large range of the decay half-life of heavy elements, propelled Gamow to star status in the Soviet Union, resulting in him becoming a member of the Academy of Sciences there at just 28 years old.

Later in life, Gamow became interested in how DNA and RNA encoded sequences used by cells to synthesize proteins and worked with James Watson (co-discoverer of the famous double helix structure of DNA). He approached this from a mathematical viewpoint, using combinatorics to deduce how a short series of bases in DNA could encode a specific amino acid. Although Gamow's model was wrong, it did inspire others to find the correct encoding.

Gamow also was a successful writer of popular science. His series of books with a lead character called Mr. Tompkins led non-scientists into the world of the atom, explaining molecular biology and quantum mechanics among other topics. Gamow was a creative and colorful character, a true renaissance man of many talents.

Some authors have argued that our universe would be very different, perhaps even uninhabitable, if a stable nucleus with mass 5 existed. In that fictional case, there would be less hydrogen after the Big Bang because more nucleons could be absorbed on helium (^{4}He). This could change the fusion reactions in stars (see Section 5.1). However, new theoretical calculations of Big Bang nucleosynthesis, modified for an imaginary world that includes stable nuclei having mass 5 and mass 8 (Coc et al., 2012) and doesn't significantly alter the nucleosynthesis. In other words, it's not as big a problem as previously stated. Your job as chef of the elements is slightly easier.

Forming heavier nuclei is difficult due to simple thermodynamics (Turner, 2022). It requires more collision energy between nuclei with more than one proton to overcome electrical repulsion. Recall that like charges repel. So, the more protons in the collision, the more the two nuclei repel each other. As the universe expanded after the Big Bang, it cooled rapidly. The lower temperatures mean that the collisions had less energy. Hence, the production of heavy nuclei from the Big Bang is stifled.

So where did the carbon, oxygen, and other heavier elements that make up cells in your body come from? The answer, as you may have heard, is that we're all made from stardust. These elements were forged in the center of stars (and some from explosive events such as supernovae). The story of how this happens is quite interesting and points to the delicate balances in nature that make life possible.

5.1 Why Stars Shine

A star is formed when a cloud of hydrogen gas collapses in upon itself under the force of gravity. After the Big Bang, there was a long period of time when the universe was just a swirling bunch of gas, about 74% hydrogen, 25% helium (by mass), and 1% other nuclei (mostly deuterium and lithium) (Peebles, 1966). From laboratory experiments, we know that when an ideal gas is compressed, it gets hotter. The same happens to a newly forming star. The energy to heat up the gas comes from the gravitational energy (NASA, 2024). In the case of a star that has the mass of our sun, this energy is enough to heat the gas to a temperature of several million degrees. This high temperature is needed before the process of nuclear fusion can get started. For a smaller gas ball, say the size of Jupiter, the temperature doesn't get high enough to ignite fusion.

Another important thing happens when nuclear fusion starts. The fusion heats the star from within, giving the gas an outward pressure that counters the force of gravity. The delicate balance between the outward pressure and the inward gravitational force keeps the star stable. Without the heat from fusion, the star would just continue to collapse. For heavier stars, several times the mass of our sun, the collapse would end up as a black hole. Obviously, a universe filled with so many black holes would not be a nice place to live! The heating from fusion is needed to allow time for life to form.

A more massive star will have a higher temperature (and hence more outward pressure) to balance the higher gravity. That higher temperature comes from more fusion to supply that energy. The result that *heavier stars use up their nuclear fuel faster*. In other words, the bigger the star, the shorter its life span. Our sun was formed about 4.5 billion years ago and will shine for at least another four billion years. If our sun had twice the mass, then it would already be burned out by now. The sun is nicely balanced, having a small enough mass to last long enough for life to develop, but having large enough mass to reach the temperature where fusion can start. Of course, the sun is not so unique. There are billions of stars in our galaxy that are like the sun.

Getting back to the formation of heavier elements, the process to combine two deuterium nuclei to get to helium (^{4}He) is a bit more involved (see Figure 5.2) than for the Big Bang (Figure 5.1). The difference is because there are almost no free neutrons in a star like our sun, which is made mostly of hydrogen. The fusion process in a star instead starts with two protons combining, and one of the protons converting to a neutron (via the weak nuclear force, which emits a positron and a neutrino) (Martin, 2006). At each stage of the process, the nuclear reactions give off energy, shown by the particles that exit left or right off the reactions in Figure 5.2. The energy of those particles is absorbed as heat, keeping the star hot. Although the reaction chain shown in Figure 5.2 might seem complicated, it was first worked out in the late 1930s by Hans Bethe (Bethe, 1939), only a few years after the neutron was discovered. For this work, he was awarded the Nobel Prize in Physics in 1967, nearly thirty years later (see Box 5.2). Hans was quite the nuclear chef.

Sidebar 5.2 Hans Bethe

Born in Strasbourg in 1906, Hans Bethe made important contributions to the development of the atomic bomb (the kind dropped on Nagasaki) and the hydrogen bomb (a much more powerful bomb, as seen in test detonations in the Pacific Ocean). However, in the scientific world, he is known mostly for his contributions to theoretical physics, such as his theory of stellar nucleosynthesis, which earned him the 1957 Nobel Prize in Physics.

continued

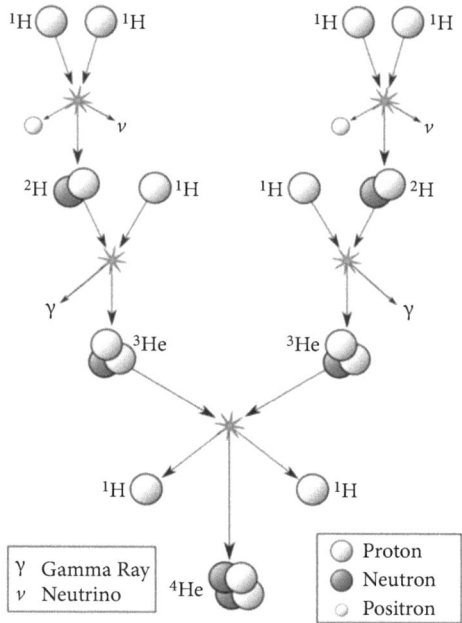

Figure 5.2 Fusion reactions in the sun. At the top, two hydrogen atoms combine (via the weak-nuclear force) to form deuterium. Adding another proton gives ^{3}He (middle). Combining two ^{3}He nuclei gives helium (^{4}He) plus two protons (^{1}H).
Reproduced from Borb (2006). Via Wikipedia Commons (CC BY-SA 3.0).

Sidebar 5.2 *continued*

Hans Bethe is one of many physicists of Jewish origin who emigrated to the United States during the early 1930s. The success of building an atomic bomb, as portrayed in the film *Oppenheimer*, is largely due to the many contributions by Jewish scientists who came to America to evade the Nazi regime. While in Germany, Bethe didn't work on nuclear physics, but quickly became one of the experts in this fledgling field after moving to Cornell University, which was hiring new faculty for this purpose.

During a conference at George Washington University in 1938, Bethe became interested in how stars shine. George Gamow (see Box 5.1) and colleagues had earlier proposed that the sun's energy came from the fusion of protons (top of Figure 5.2). However, this hypothesis did not explain how elements heavier than deuterium occurred. By the end of the conference, Bethe and others there had come up with a series of nuclear reactions that could also occur in the sun, leading to light elements like helium (bottom of Figure 5.2), lithium, and beryllium being produced.

Later, back at Cornell, Bethe came up with a process called the CNO cycle (nuclear reactions on carbon, nitrogen, and oxygen) that could occur in some stars—those more massive than our sun, where interior temperatures are higher. It was for this work that he won the Nobel Prize. It would not be until 2002, more than a half-century later, that Bethe's ideas could be experimentally confirmed by measurements of neutrinos emitted from the sun.

After the Manhattan Project ended, Bethe continued to make major contributions to theoretical physics, including topics such as how supernovae explode. He also solved a problem with energy states of the hydrogen atom, called the Lamb shift (discovered by Willis Lamb and his graduate student), which was hugely influential for quantum theory. It is said that Bethe produced influential papers for seventy years, from in his twenties to his nineties. Few scientists have had such productive careers.

If we look at stars in galaxies far away, we see those stars as they were billions of years ago. The reason we can look back in time is that those far-off galaxies are billions of light-years away, meaning it takes light (even at its phenomenally fast speed) billions of years to cross the void. By looking at the **spectrum of light** from these far-away galaxies, we see stars as they were forming soon after the Big Bang. The spectrum tells us about the atoms present in the star, since each element gives off (and absorbs) unique colors of light, essentially giving a "fingerprint" of that element. For example, if you look at a stained-glass window, the colors of glass are made by adding certain elements to the glass. For example, adding cobalt to glass gives it a distinct blue-violet color. Similarly, by looking at the spectrum of stars, astronomers can tell which elements are present.

Astronomers have observed stars that are tens of billions of light-years away, which formed soon after the Big Bang, and are made of about 74% hydrogen, with 25% helium, and less than 1% heavier elements (or "metals" as they are called by astronomers). As noted earlier, this is exactly the right mix of elements predicted from Big Bang nucleosynthesis. Those early, metal-poor stars get almost all their heat from hydrogen fusion.

Our sun is different from those early stars. Looking at the spectrum from the sun using precision instruments, its chemical makeup shows it has (in percentage of total mass) 71% hydrogen, 27% helium, 1% oxygen, 0.4% carbon, and 0.6% other elements (NASA, 2024). This is a strong indication that the sun is a second- or third-generation star, meaning that it was formed billions of years after the Big Bang, born out of the dust clouds spewed out from a previous exploding star. But where are all of the elements between helium and carbon/oxygen? Shouldn't we expect to see elements such as lithium, beryllium, boron, and so on? The answer lies in the nuclear physics.

Since a stable nucleus with five nucleons doesn't exist, combining a proton with helium isn't possible. You might think that two helium nuclei could undergo fusion, but the nucleus formed (^{8}Be, with four protons and four neutrons) is unstable, lasting only about a billionth of a billionth of a second (or 10^{-18} seconds) before decaying back into two helium nuclei. However, if we could get three helium nuclei to fuse all at once, then carbon (^{12}C) could be made, which is stable. You might think the chances of this happening are very small, and you'd be right. However, a star is a very big laboratory with "zillions" of helium atoms. If you take a very small probability and multiply it by a very large number of collisions, then it's likely to happen. This idea was first quantified by physicist Willie Fowler (see Box 5.3), who was awarded the 1983 Nobel Prize in Physics.

Box 5.3 Willie Fowler

William Fowler, born in 1911, grew up in Ohio and received a degree in engineering physics from Ohio State University. That was at the time of the Great Depression, and jobs were scarce, but Fowler got accepted for graduate school at Caltech doing research at the newly formed Kellogg Radiation Laboratory. Originally founded as a center for cancer radiation therapy by W. K. Kellogg (of breakfast cereal fame), it was soon transformed to do basic research in nuclear physics, where Fowler did the experiments that won him the 1983 Nobel Prize in Physics.

Known to his friends and colleagues as Willie, he had an engaging personality and sparked many animated discussions. In his address to the Nobel Prize committee (Fowler, 1983), he quotes Mark Twain's 1883 memoir, *Life on the Mississippi*, saying, "There is something fascinating about science. One gets such wholesale returns of conjecture out of such a trifling investment in fact." This expresses the attitude of many experimental and theoretical physicists, who work long hours trying to understand what Nature is telling us from laboratory measurements.

Fowler's most famous paper was co-authored with Margaret and Geoffrey Burbidge, along with Fred Hoyle (see Box 5.4). Often quoted as B^{2}FH, after the initials of the authors' last names, this laid out the case for the origin of how the known elements were produced and, more importantly, the abundance of those elements. This review paper described several chains of nuclear reactions, with names such as r-process (rapid production) and s-process (slow production) that are still hot topics for today's research at laboratories such as the new Facility for Rare Isotope Beams (FRIB) that opened in 2022 at Michigan State University. Sadly, none of Fowler's co-authors on this paper received the Nobel Prize, but all of them contributed substantially to this landmark research.

Fowler believed that basic research was hard work but also fun. There are few things as exhilarating as spending hour upon hour working toward a goal, puzzling over how to make sense of a measurement, and then having the light bulb turn on that illuminates what Nature is doing. This happened for Willie Fowler as he measured, for the first time ever, some of the nuclear reactions that hinted at how the sun shines and how the light elements are produced. Imagine being one of the first people to figure out that nuclear reactions can explain how carbon (and other elements necessary for life to exist) are made by our universe.

The actual process that happens in a star is shown in Figure 5.3. First, two helium atoms fuse to make ^{8}Be, which emits some energy (as a gamma ray) and sticks around just long enough for another helium atom to come by and get absorbed, and then it emits more energy (another gamma ray). This sounds like a straightforward process, but the details are more subtle than you might think, as explained below.

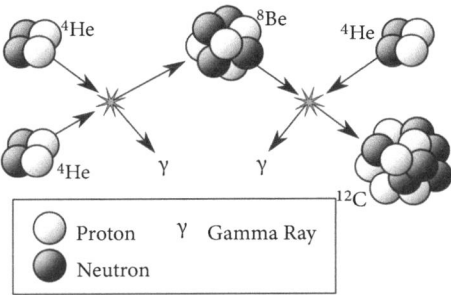

Figure 5.3 The fusion process in a star to form carbon. First, two helium nuclei (^{4}He) combine to form a metastable state of beryllium (^{8}Be), which combines with a third helium nucleus to form ^{12}C.
Reproduced from Borb (2006). Via Wikimedia Commons (CC BY-SA 3.0).

Atoms (nuclei with electrons orbiting) and nuclei (only the protons and neutrons) both obey the laws of quantum physics. In the case of atoms, the electrons can absorb only certain colors (wavelengths) of light because the electrons must orbit in specified quantum energy states (see Chapter 2). Unless the energy of the light beam matches with the energy difference between quantum states, the light goes straight through it. As noted above, glass is transparent to visible light. This is because there is a mismatch between the energy of electron orbits in glass (its quantum states) and the energy carried by visible light. As a result, visible light sails through glass unimpeded.

Similarly, nuclei can absorb (or emit) energy only at certain wavelengths, because the protons (or neutrons) also orbit in their quantum energy states. Because the nucleus is much smaller than the orbital radius for electrons, it takes more energy to transition to a nuclear quantum state.[2] But here's the catch. If the energy level isn't just right, then nuclear fusion is suppressed. The main takeaway message is this: the energetics of nuclear collisions must be "just right"; otherwise, the nuclear reaction doesn't happen.

Here we encounter an almost miraculous coincidence. It turns out that a quantum level of ^{12}C is almost exactly matched to the combined mass of the two nuclei, ^{8}Be and ^{4}He (Dunbar et al., 1953) (see Box 5.4). Without this nearly perfect mass-energy balance, then the reactions shown in Figure 5.3 that produce carbon nuclei couldn't happen. Of course, without carbon there would be no life on Earth.

Box 5.4 Fred Hoyle and the Carbon Prediction

To give you a sense of how perfectly balanced the nuclear reactions are to make carbon, here are the first few quantum energy states of ^{12}C (in units of MeV): 0.0, 4.43, 7.66, and 9.63. The value 0.0 is called the "ground state," which represents a carbon nucleus at rest. If a gamma ray with energy 2.0 MeV comes along, it will sail right through carbon, since it doesn't have the right energy to be absorbed. However, a gamma ray with an energy of 4.43 MeV would be absorbed by carbon. The spacing of the quantum energy states is dictated by the value of the strong-nuclear force.

Looking at the exact masses for the nuclei that fuse together to form carbon, it turns out that there is a slight imbalance. The combined mass of ^{8}Be and ^{4}He is slightly heavier than ^{12}C. When converting this mass excess to energy (using Einstein's equation $E=mc^2$), the difference is 7.65 MeV. This mass difference could have been any number, but it just happens to be nearly equal the second quantum state of carbon. Adding a bit of kinetic energy from the collision of atoms gets us an exact match, allowing those two nuclei to fuse into carbon. The excess energy, 7.66 MeV, is then given off by the carbon nucleus as a gamma ray, which is absorbed in the surrounding matter, thus heating the star.

The history of the discovery of this quantum state in ^{12}C is fascinating, and it's a great example of the predictive power of physics. Back in 1954, not much was known about the quantum states of nuclei. Fred Hoyle, an astronomer, knew that there was a lot of carbon in stars (like the sun) and had been working with Willie Fowler on how carbon was produced in the universe. They had calculated that very little carbon

---

[2] This is due to the uncertainty principle. When a particle is confined to a smaller space, the uncertainty principle forces that particle to have a larger momentum (or energy).

was made by the Big Bang (Fowler, 1983), so they looked to stellar fusion as a way to produce it.

Using the precise masses of ^{8}Be and ^{4}He, which had only just been measured, Hoyle realized that the nuclear reaction could not happen unless there was a quantum state in ^{12}C at about 7.65 MeV (Hoyle, 1954). At that time, only the first quantum level, at 4.43 MeV, was known. After convincing the nuclear experimentalists at Caltech's Kellogg Laboratory to do more measurements at higher energy, they found the quantum state at 7.66 MeV.

This was perhaps the first precise prediction of a quantum state in a nucleus, because knowledge of the nuclear force wasn't good enough back then to make such a prediction. This quantum state in ^{12}C is forever known as the Hoyle state. By the time the 1983 Nobel Prize was awarded for this work (see Box 5.3), Fred Hoyle had unfortunately passed away (the Nobel Prize can only be awarded to living recipients).

So, nature has this exquisite balance between the masses and quantum energy states of the nuclei involved in the nucleosynthesis of carbon. If the nuclear force were just a few percent weaker, there would be a mismatch in the energy needed for the process shown in Figure 5.3, and carbon formation would be highly suppressed in stars. Similarly, if the nuclear force were just a few percent stronger, again there would be a mismatch, and the star could not produce a large amount of carbon (Adams and Grohs, 2017).

This is the second big coincidence that we've encountered from the nuclear force. The first one was discussed in Chapter 4, where the nuclear force is too weak to bind two protons but is just strong enough to bind a neutron and a proton. In either case, the universe would be very different, and life as we know it wouldn't exist. That itself is a coincidence of **"fine-tuning"** of the nuclear force (Barnes, 2012). Now, we see that again the nuclear force must be fine-tuned *to exactly the same value*; otherwise, the process in Figure 5.3 wouldn't have the right energy to form carbon. Two such coincidences, each balanced "just right" for life to exist, might make you take pause.

To give a sense of how unusual the situation is for carbon, if we now try to fuse together helium and carbon, to form oxygen (^{16}O, with eight protons and eight neutrons), that reaction doesn't go well. The quantum resonances in oxygen do not match up well with the mass differences for these nuclei. As a result, this pathway is partly blocked, creating a bottleneck in the pathway to heavier nuclei.[3] The same is true for adding helium to oxygen, as there is

---

[3] It's still possible to manufacture oxygen in a star via fusion reactions, but the reaction rate is much reduced because of the mismatch of the quantum resonance. Instead, there is another pathway to produce oxygen, but the nuclear physics to do this is more complicated (called the CNO cycle).

no matching quantum resonance in neon (^{20}Ne, with ten protons and ten neutrons). So, we see that it really is coincidental that it just happens to work out for the case of making carbon.

5.2 Big Stars, Little Stars

A story of stellar evolution is beyond the scope of this book, but there are some key points that are useful for later chapters. One of the most important things to know is that the fate of a star depends on its mass. Some stars will explode, and others won't.

For stars like the sun, which has exactly 1 solar mass (1 M_{Sun}) by definition, the primary heat source is hydrogen fusion (Figure 5.2). As mentioned above, the sun takes about ten billion years to burn its hydrogen fuel. When the hydrogen is almost spent, the helium core of a star starts to compress, raising its temperature to where it can start to burn helium (Figure 5.3). The heat radiated from helium burning will cause the outer envelope of hydrogen to expand, becoming a "**red giant**" star. This won't happen to our sun until about four billion years from now, but when it does, it will engulf Earth as it expands, ending all life on Earth.

After time, a red giant may accumulate carbon at its core. If there isn't enough heat from helium fusion to keep the star stable, its core will collapse. Then the red giant ejects its outer layers as a planetary nebula (NASA, 2024), leaving a **white dwarf** star that is about the size of Earth but made mostly from carbon and with a fractional amount of oxygen and nitrogen (which are produced when lighter nuclei fuse with carbon). The fractional amount depends on the mass of the star. Figure 5.4 shows this evolution.

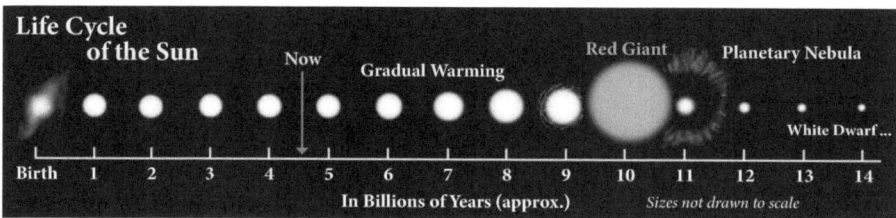

Figure 5.4 Evolution of the sun, from birth as a cloud of gas and dust, to the short red giant phase, ending as a small white dwarf star. (The sizes shown are not to scale; see note in lower right.)

The fate of the sun is shared with other stars, ranging from about half of a solar mass up to stars with 8 solar masses ($8M_{Sun}$). For stars less than $0.4M_{Sun}$, they don't get hot enough to burn helium, and eventually exhaust their hydrogen fuel. In fact, they burn hydrogen so slowly that it takes a very long time (longer than the current age of the universe) for them to burn out.

For stars with mass greater than $8M_{Sun}$, these get hot enough to burn carbon, forming heavier and heavier nuclei, until it reaches iron. Iron is the endpoint to fusion, because heavier nuclei have less binding energy (so iron fusion doesn't release energy).[4] At this point, the core of these massive stars collapses, and the weight is so great that even the electron obits are crushed, so that all the nuclei merge (NASA, 2024). The result is a supernova, leaving either a **neutron star** (a star made entirely from neutrons and held together by gravity) or a black hole. The latter occurs for supermassive stars that started with more than 25 solar masses ($25M_{Sun}$). The neutron stars are of special interest, as we'll see in a later section, but first let's explore the supernovae explosions.

5.3 Supernovae

There are several types of **supernovae**, called type I and type II, but first we'll focus on the collapse of dying stars described above (type II). The types of supernovae are categorized by the spectrum of light (the distribution of colors) that they give off (NASA, 2011). Type II supernovae have colors that show lots of hydrogen was present, whereas type I supernovae are missing those characteristics. We expect that hydrogen is present in the outer layers of an exploding star, and that's one way that astronomers know to identify the type II spectrum as an explosive event.

The physics of how a massive star explodes is still fraught with uncertainties. Some aspects, however, are reasonably certain. As the star's core collapses when its fuel is used up, this leads to an iron core. In some cases, the core collapses, and the electrons are captured by protons (via the weak nuclear force), leading to a neutron star. What happens next depends on several factors, including how massive the star is, the fraction of "metals" present, and how "stiff" the neutron star is. The star may or may not explode, according to advanced supercomputer calculations (Burrows, 2019).

---

[4] The iron nucleus is at a balance point, where the repulsive force between the like-charged protons and the attractive nuclear force gives it the most binding energy per nucleon.

In previous years, the conventional wisdom was that the matter above the core collapse would fall onto the neutron star and rebound, creating a shock wave of matter moving outward from the center. This picture was based on theoretical models that assumed spherical symmetry, meaning that it looked the same in all directions outward from the core. More sophisticated computer simulations, done in three dimensions, show that there is a substantial departure from symmetry, meaning that the physics of supernovae explosions has much more asymmetry (and hence more complexity) than initially thought (Burrows, 2019). As computers get faster and the physics used to model it becomes more sophisticated, the result is that there is no "one-size-fits-all" explanation of type II supernovae.

The one thing we know is that massive stars certainly do explode, and these supernovae spew out heavy nuclei into the surrounding space. We see the results in the sky, with many nebulae observed, which are the debris fields left behind. These clouds of dust (the nebulae) are rich in elements ranging from carbon to iron and smaller amounts of heavier nuclei. As stars orbit around the center of the galaxy,[5] the dust clouds from many supernovae become intermixed. Over time, these dust clouds give way to gravity and start to collapse inward, forming new stars and planets. This is much like the process that formed our solar system.

Figure 5.5 shows a magnified image of the edge of a gas cloud, taken by the Hubble Space Telescope (HST), showing a region where new stars are being born. This now-famous picture has been dubbed "**pillars of creation**," although some scientists think that's a lofty name for such a common process in the galaxy. The recirculation of dust from previous supernova explosions has been happening for billions of years. NASA estimates that a few new stars per year are formed in the Milky Way galaxy alone (Evans, Kim, and Ostriker, 2022). After a billion years, that's a lot of stars!

Astronomer Carl Sagan famously said, "The nitrogen in our DNA, the calcium in our teeth, the iron in our blood, the carbon in our apple pies were made in the interiors of collapsing stars. We are made of star-stuff" (Sagan, 1980). These words are a wonderful summary, but I would add that these elements (carbon, nitrogen, and the heavier nuclei) are only here because of a delicate balancing act in the nucleosynthesis process. Change the nuclear force slightly, and nucleosynthesis breaks down. Our universe has just the right balance for life, and at present nobody knows why.

---

[5] Our sun goes in a 250-million-year journey around the center of the Milky Way.

Figure 5.5 HST image of the Eagle Nebula, showing part of a massive cloud of gas and dust where new stars are forming.

Reproduced from NASA, Hester and Scowen (2003). Via Wikimedia Commons (public domain).

5.4 Neutron Stars

One question that has plagued nuclear scientists for years is how the heaviest elements, such as gold, lead, and uranium, could be produced in supernova explosions. It's known that heavier elements can be made through absorption of neutrons, and there are a lot of neutrons flying around when a supernova explodes. However, sophisticated computer calculations including all the nuclear reactions (either measured or modeled by nuclear theory) showed only small amounts of the heaviest nuclei being produced by supernovae (Clayton, 2008). Either the nuclear physics is wrong, or there is some other process to form the heaviest nuclei.

There has been a paradigm shift in the field of nuclear astrophysics since the discovery of two neutron stars merging, captured by gravitational waves (Abbott et al., 2017a), along with the optical afterglow observed by several telescopes. Before this event, some theoretical physicists had suggested that the heaviest elements could be produced when two neutron stars merged, throwing off neutron-rich matter in the process. While other scientists doubted this idea, it turns out that they were right.

Systems of two neutrons stars orbiting each other had been known for decades, but what happened after their orbits spiraled inward (due to emission of gravitational waves) or the two neutron stars merged was uncertain. Computer models, based on Einstein's theory of gravity and models of the nuclear equation of state (how neutrons and protons behave under extreme pressures), calculated that the neutron stars would stretch as they approached each other (Read et al., 2009) and then fling out large masses of neutrons before merging (Abbott et al., 2017b). These clumps of neutrons could then undergo rapid radioactive decay, forming heavy nuclei. While this appeared fine in theory, there were no experimental constraints. For example, nuclear physicists haven't produced clumps of pure neutron matter in the laboratory (all nuclei have a mix of neutrons and protons), and so no one could be sure what nuclei would be produced by a merger of two neutron stars.

The discovery by the **LIGO** (Laser Interferometry Gravitational Wave Observatory) Collaboration in 2017 (labeled GW170817) changed everything. LIGO not only saw the signature of two neutron stars merging in their gravitational wave detectors but they also could locate a rough position in the sky of where it happened (Abbott et al., 2017a). Astronomers quickly turned their telescopes to the right coordinates and saw a bright spot that hadn't been there before. Measurements of the afterglow, which is caused by the light given off when nuclei are being formed, agreed well with the predictions from computer simulations of the merger of two neutron stars. In particular, the light seen in telescopes changed from bluish to reddish over the following days in the way expected if heavy nuclei (such as gold and uranium) were being produced (Abbott et al., 2017b). The changing light spectrum was also very different from that seen for the various types of supernovae.

The idea that the heavy elements here on Earth were formed by neutron star mergers, and not from supernovae, is profound. It impacts our daily life, although the connection may not be obvious. For example, the dynamo that drives the liquid magma deep in the Earth is driven by the presence of heavy elements. The heat from radioactive decay of uranium and thorium keeps the magma hot enough to stay liquid and circulating, much like heat in a convection oven. In turn, the convection of liquid rock below the Earth's surface produces a magnetic field, which makes a compass needle point north. This magnetic field is important because it *deflects harmful cosmic rays from space*. **Cosmic rays** are a threat to life, much like ultraviolet light from the sun is harmful to your skin, except that cosmic rays are many times more harmful. Life on Earth would be negatively impacted if not for the shielding effect of our magnetic field (Erdmann, Kmita, Kosicki, and Kaczmarek, 2021). This magnetic "shield" depends on there being radioactive nuclei in Earth's interior, and those nuclei were likely created from mergers of neutron stars.

The gold in a wedding ring and the uranium in the Earth are likely the product of a phenomena (neutron stars) that is just starting to be understood. At present, there is no other reasonable explanation for the source of very heavy nuclei (at least in the quantities found in our solar system) except for neutron star mergers (see Box 5.5). These objects, once thought to be of interest only to academics, may be necessary for human evolution. Life below the ocean's surface would not be greatly impacted by cosmic rays; however, could life on land have developed without Earth's magnetic shield?

The existence of neutron stars depends on a balance between the nuclear force and the force of gravity, which determines the possible mass and radius that a neutron star can have (Lattimer and Prakash, 2001). However, much more research on neutron stars is needed before we can say more about this balance point of Nature (see Figure 5.6).

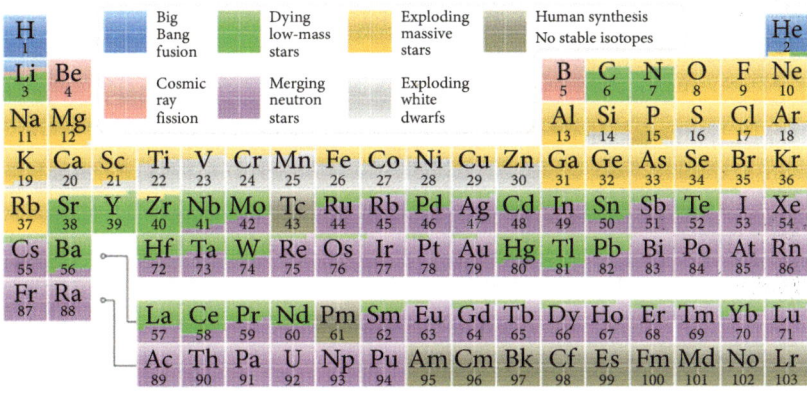

Figure 5.6 Periodic table of the elements, with color coding of the source of the element (as shown in the key at the top).
Reproduced from Cmglee (2014). Via Wikimedia Commons (CC BY-SA 3.0).

Box 5.5 Reticulum II

One piece of evidence that merging neutrons stars are responsible for the existence of very heavy elements (like gold or europium) comes from astronomical observations of a faint dwarf galaxy that orbits the Milky Way, called Reticulum II. It is about

continued

Box 5.5 *continued*

100,000 light-years from Earth, having a smallish cluster of stars. (The name comes from Latin for "small net.") Reticulum II was found by an astronomy project called the Dark Energy Survey, which scanned the sky to catalogue all the stars and their spectrum of light.

It has been known for many years that there are several dwarf galaxies orbiting the Milky Way, but the stars in Reticulum II were cataloged only recently (2014–2015). As mentioned earlier, the spectrum of light from stars carries information about the elements that make up a star. Curiously, many of the stars in Reticulum II showed a spectrum that contained very heavy elements. This dwarf galaxy is too far away to have gotten contaminated by these elements from the Milky Way. Either there were a whole bunch of supernovae among this small collection of stars, which seemed highly unlikely, or a single merger of two neutron stars could explain it.

Although it is a rare phenomenon, the merging to two neutron stars has a unique signature, producing a large amount of very heavy elements. These heavy nuclei get flung outward by the centrifugal force as the neutron stars are first elongated and then torn apart by gravity as they spiral toward each other. So, in a small galaxy like Reticulum II, the stars that form from collapsing dust clouds will inherit the heavy elements, along with their telltale spectra of light that they can emit (or absorb). By looking at the spectrum of stars with high resolution, astronomers can tell what elements have been captured by a star. It is estimated that about 72% of the stars found in Reticulum II have spectra that show significant amounts of heavy elements (also called r-process elements) (Simon et al., 2023).

The idea that Reticulum II had a rare event, where two neutron stars merged, is still controversial. More research is needed to calculate the amounts of heavy elements formed from merging neutron stars, and more gravitational wave observations of binary neutron star mergers will be forthcoming. Nonetheless, this idea is currently the simplest explanation that fits the astronomical observations of Reticulum II. Sometimes Nature is kind, and this dwarf galaxy is a lucky find.

A chart of the elements is shown in Figure 5.5, where the shading of the boxes shows the portion of each element that comes from either the Big Bang, supernovae, or neutron star mergers. In addition, some elements do not occur in Nature, such as technetium (Tc), because of their short half-life for radioactive decay. The proportions (shadings) of the sources may change as more research is done on the process of producing nuclei in both supernovae and neutron star mergers, but the overall picture at present is that most of the heavy elements come from neutron star mergers. This is a recent break-through in our understanding of where the heavy elements come from (Chen, Vitalie, and Foucart, 2021).

References

Abbott, B. P. et al. (The LIGO-Virgo Collaborations) (2017a) "GW170817: Obser-vation of gravitational waves from a binary neutron star inspiral," *Phys. Rev. Lett.*, 119, pp. 161101.

Abbott, B. P. et al. (The LIGO-Virgo Collaborations) (2017b) *ApJL.*, 850, L39.

Adams, F. C. and Grohs, E. (2017) "Stellar helium burning in other universes: A solution to the triple alpha fine-tuning problem," *Astropart. Phys.*, 87, pp. 40–54.

Alpher, R. A., Bethe, H., and Gamow, G. (1948) "The Origin of Chemical Ele-ments," *Phys. Rev.*, 73, pp. 803–804. See also: R. A. Alpher and R. C. Herman (1950) *Rev. Mod. Phys.*, 22, pp. 153.

Barnes, L. A. (2012) "The Fine-Tuning of the Universe for Intelligent Life," *Pub. Astro. Soc. Aus.*, 29, pp. 529–564.

Bethe, H. A. (1939) "Energy Production in Stars," *Phys. Rev.* 55, p. 434.

Burrows, A., Radice, D., and Vartanyan, D. (2019) "Three-dimensional supernova explosion simulations of 9-, 10-, 11-, 12-, and 13-M$\odot$ stars," *MNRAS*, 485, pp. 3153–3168.

Chen, H.-Y., Vitale, S., and Foucart, F. (2021) "The Relative Contribution to Heavy Metals Production from Binary Neutron Star Mergers and Neutron Star–Black Hole Mergers," *ApJL.*, 920, L3.

Clayton, D. D. (2008) "Fred Hoyle, primary nucleosynthesis and radioactivity," *New Astronomy Reviews* 52, pp. 360–363.

Coc, A., Descouvemont, P., Olive, K. A., Uzan, J.-P., and Vangioni, E. (2012) "Variation of fundamental constants and the role of A = 5 and A = 8 nuclei on primordial nucleosynthesis," *Phys. Rev. D*, 86, p. 043529.

Dunbar, D. N. F. et al. (1953) "The 7.68-Mev State in C-12," *Phys. Rev.*, 92, p. 649.

Erdmann, W., Kmita, H., Kosicki, J. Z., and Kaczmarek, L. (2021) "How the Geomag-netic Field Influences Life on Earth – An Integrated Approach to Geomagnetobi-ology," *Astrobiology* 51, pp. 231–257.

Evans, N. J., Kim, J-G., and Ostriker E. C. (2022) "Slow Star Formation in the Milky Way: Theory Meets Observations," *Astrophysical Journal Letters*, 929, L18.

Fowler, W. A. (1983) Nobel Lecture, December 8.

Hoyle, F. (1954) "On Nuclear Reactions Occuring in Very Hot STARS.I. The Syn-thesis of Elements from Carbon to Nickel," *Astrophysics Journal Supplement Series*, 1, p. 12.

Lattimer, J. M. and Prakash, M. (2001) "Neutron Star Structure and the Equation of State," *ApJ.*, 550, pp. 426–442.

Lauritsen, T. and Ajzenberg-Selove, F. (1966) "Energy levels of light nuclei (VII). A = 5–10," *Nucl. Phys.*, 78, p. 1.

Martin B. R. (2006) *Nuclear and particle physics: An introduction.* West Sussex, UK: Wiley. Chapter 8, Section 8.2.2.

NASA, (2011). Available at: https://imagine.gsfc.nasa.gov/science/objects/superno vae2.html.

NASA, (2024). Available at: https://science.nasa.gov/universe/stars/.

Peebles P. J .E. (1966) "Primeval Helium Abundance and the Primeval Fireball," *Phys. Rev. Lett.*, 16, pp. 410–413.

Read, J. S. et al. (2009) "Measuring the neutron star equation of state with gravitational wave observations," *Phys. Rev. D*, 79, pp. 124033.

Sagan, C. (1980) *Cosmos* TV series.

Simon, J. D. et al. (2023) "Timing the r-process Enrichment of the Ultra-faint Dwarf Galaxy Reticulum II," *ApJ.* 944, pp. 43.

Turner, M. S. (2022) "Understanding BBN: The physics and its history." Available at: arXiv:2111.14254.

Zoroddu, M. A., Aaseth, J., Crisponi, G., Medici, S., Peana, M., and Nurchi, V. M. (2019) "The essential metals for humans: A brief overview," *Journal of Inorganic Biochemistry*, 195, pp. 120–129.

Chapter 6
The Left Hand of Life

All of us who study the origin of life find that the more we look into it, the more we feel it is too complex to have evolved anywhere. We all believe as an article of faith that life evolved from dead matter on this planet. It is just that life's complexity is so great, it is hard for us to imagine that it did.

—Harold Urey

Perhaps the biggest uncertainty in the origin of life on Earth is not the structure of the universe. Rather, it is how life got started. Of course, a prerequisite for life is that the balance of physical laws in the universe are just right, as described in the previous chapters. Once the conditions are ripe, with elements such as carbon, nitrogen, and oxygen that can form molecules of **amino acids** (the building blocks of proteins) and **nucleotides** (the building blocks of RNA and DNA), the chemistry of life becomes possible. But just because it's possible doesn't mean that it's inevitable.

Due to advances in molecular biology, we now know the structure of many molecules that are necessary for life. For example, we know the chemical structure of many proteins that carry out the daily cellular functions that keep us alive. We also know the mechanisms by which DNA builds these protein molecules. These concepts took centuries of research to uncover but are now standard textbook material. Yet when life started, billions of years ago, the chemistry of life must have been much simpler (see Box 6.1). Can we take our current knowledge and extrapolate backward in time to the first molecules that could reproduce copies of themselves?

Box 6.1 Life on Early Earth

Measurements of radioactive decay in the oldest rocks (including meteorites and moon rocks) dates the age of Earth at about 4.5 billion years old (4.5 Ga).[1] Similar dating techniques put the age of the oldest fossils (called stromatolites, built from cyanobacteria—simple organisms that can absorb sunlight) at about 3.5 Ga. Indirect

continued

Nature's Balancing Act. Ken Hicks, Oxford University Press. © Oxford University Press (2025).
DOI: 10.1093/9780197771471.003.0006

Box 6.1 *continued*

---

evidence of life (based on graphite inside mineral grains) suggests life started over 3.7 Ga ago. So, it appears that life appeared on Earth almost as soon as it possibly could.

In models of the early solar system, the Earth was a violent place. It had molten rock spewing up from below and large meteorites constantly landing, causing massive craters. Current thinking is that the Earth started to settle down enough to have oceans about 4.2 Ga ago. Only then could the building blocks of life (amino acids and nucleotides) mix to form more complex molecules.

Conventional wisdom says that a molecule as complex as DNA, with its double-helix backbone, could not have come together by random chance. But if you put nucleotides (also called bases) into a test tube along with ribose (a simple type of sugar), some of these will bond together to form short sections of RNA (ribonucleic acid). RNA looks much like half of a DNA molecule, with only a single helix, but is more fragile. It easily breaks apart compared with the more stable DNA. The nucleotides are attracted to the inside of the helix, where they can attach with a regular spacing.

DNA can only reproduce in the presence of a specialized enzyme (called DNA polymerase). However, short strands of RNA can form spontaneously. The problem is that the longer the strand, the more likely it will break apart somewhere along the strand due to thermal motion in water. From experimental observations, one can calculate the probability of a long strand of RNA surviving for a given period. Fold in the length of RNA necessary to have a self-replicating molecule and the time needed for that molecule to reproduce, and the chances for life to appear are astronomically small.

With research under carefully controlled conditions, using human-prepared solutions of nucleotides and sugars kept at constant temperature, it is possible to form simple RNA molecules that will reproduce themselves (see text). But simple replication is not enough. Life would need to evolve. Can life form under the natural conditions of the Earth's early oceans? Can we answer the age-old question of whether life formed spontaneously on Earth?

---

[1] Gigayears, or Ga, is the unit used by geologists and is equal to one billion years.

Currently, the simplest molecule that is known to carry reproductive information *and* catalyze organic functions is called a **ribozyme**, discovered independently[2] by Thomas Cech (Zaug and Cech, 1986) and Sidney Altman (Guerrier-Takada and Altman, 1986). A ribozyme is made from RNA but can also catalyze reactions like an enzyme (which is made of protein). In today's evolved life, enzymes are used by cells to regulate chemical reactions, such as the replication of DNA. But enzymes are complex molecules. It's likely

---

[2] Nobel Prize in Chemistry, 1989, https://www.nobelprize.org/prizes/chemistry/1989/summary.

that early forms of life used a simpler chemistry. That chemistry could be ribozymes, without any proteins. This is called the "RNA world" hypothesis.

This is a chicken-and-egg type of problem: to make enzymes, you need the information encoded by DNA/RNA, but to replicate DNA, you need enzymes. Which came first: DNA/RNA or proteins? The ribozyme is a molecule that can carry the heritable information (like for DNA) and catalyze chemical reactions, too. Also, proteins don't carry genetic coding, so they can't reproduce themselves. Hence, the RNA World hypothesis for the origin of life.

Francis Crick, co-discoverer of the structure of DNA, said, "Possibly the first 'enzyme' was an RNA molecule with RNA replicase properties" (Crick, 1968). In an ideal picture of how life got started, you would start with a ribozyme made up of, say, about a dozen nucleotides, which can reproduce itself. The reason it should have only a dozen (or so) molecules is that this primitive molecule likely formed from random collisions of nucleotides that were present on the early Earth, so the shorter the better.

In a well-known experiment conducted by Miller and Urey (see Box 6.2), under the conditions of primitive Earth with methane and ammonia in the atmosphere (see Figure 6.1), it spontaneously made simple organic

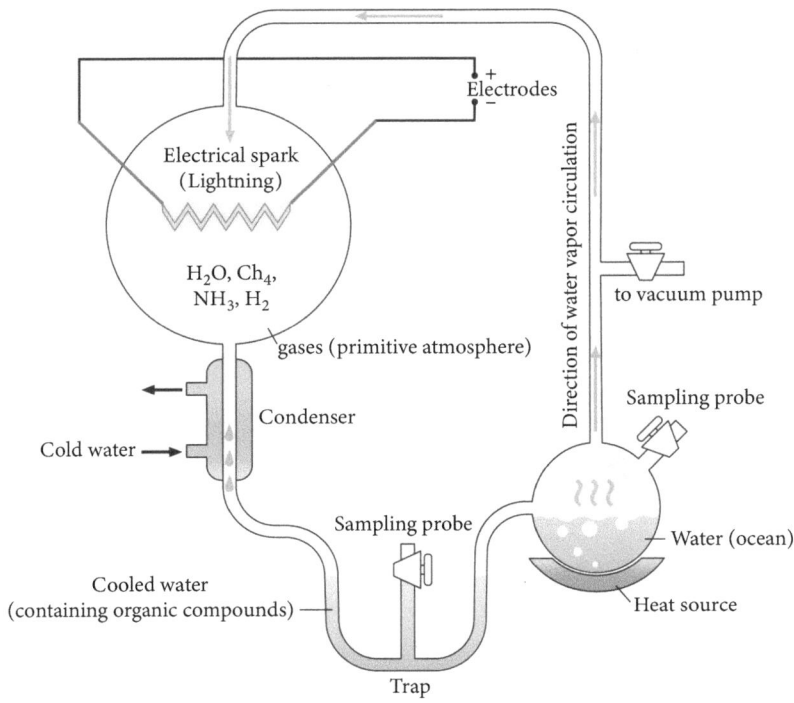

Figure 6.1 Schematic of the Miller-Urey experiment.
From Mrabet (2008). Via Wikimedia Commons (CC BY-SA 3.0).

molecules like amino acids and sugars like ribose (the sugar found in RNA). Unfortunately, more complex compounds (like a ribozyme) have not been found spontaneously in laboratory experiments that simulate the environment of the early Earth. Sometimes, one can find molecules made of a few nucleotides, but the more complex the molecule, the less likely it will form from random collisions. To build a complex molecule, nature appears to need assistance from a catalyst (such as an enzyme). But enzymes are complex molecules built from the genetic code instructions in RNA, and so we're back to the chicken-and-egg problem.

Box 6.2 The Miller-Urey Experiment

When Stanley Miller was a graduate student at the University of Chicago, he was motivated to work on the origin of organic molecules when attending a chemistry seminar by Harold Urey, winner of the 1934 Nobel Prize in Chemistry. Miller later approached Urey to work on an experiment that could mimic the conditions thought to be present in the early Earth's atmosphere. Urey was not enthusiastic since prior work on that topic had not been successful. But this did not dissuade Miller, and he eventually convinced Urey to try it.

The Miller-Urey experiment has three basic components. First is a flask of water to simulate the ocean. It is connected to a second flask above which has gasses such as ammonia, methane, and hydrogen, which simulated the early Earth's atmosphere. (There was little or no oxygen gas because there were no plants to generate the oxygen.) The third, and most important component, was electrodes to generate a spark, simulating lightning. The sparks break apart the gas molecules into ions, which can recombine to form more complex organic molecules. Indeed, they found their experiment made several amino acids (used to make proteins). A few years later, another scientist showed a similar process could make adenine, one part of a nucleotide.

Darwin speculated that life began in a "warm pond," where the organic molecules could mix and, over long periods of time, form primitive chemicals (from which life would spontaneously arise). The Miller-Urey experiment had shown that, in principle, this was possible. However, the experiment raises almost as many questions as it answers. For example, would the concentration of organic molecules be sufficient to mix, forming nucleotides and sugars, followed by longer chains of RNA? Would an alternate heat source, such as hydrothermal vents, provide better conditions than lightning? Would the early Earth have a sufficiently stable environment to provide the long time needed for organic molecules to mix? In other words, the problem of how life formed is far from over.

Many decades have passed since the Miller-Urey experiment. Many variations under different conditions have been tried. Yet science is still stuck with only being able to make simple molecules when simulating the early Earth's environment. Longer molecules can be made only when enzymes (or ribozymes, see Box 6.3) are added to the mix. The method taken by nature continues to elude even the best scientists. With more research, we may yet crack open this mystery, but for now it remains another example of a delicate balance point that could have tipped the other way, leading to no life on Earth.

One of the simplest ribozymes known (at the time of writing—this is a rapidly developing field)[3] that can copy with high fidelity (about 95% accuracy) an RNA strand of about a dozen nucleotides was made in the laboratory using the assistance of modern organic chemistry (Johnston et al., 2001). But this ribozyme is made of 189 nucleotides. In other words, it can't reproduce itself. Even then, nature would need to connect 189 simple molecules in just the right order by random collisions to make this ribozyme in the first place. Perhaps there is a simpler molecule that could reproduce exact copies of itself, with the ability to carry reproductive information as RNA. If so, though, that molecule has been lost to the past, at least for now. The bottom line is that no molecular biologist has yet found how to start with a soup of simple organic molecules (such as nucleotides mixed with simple sugars, like in the Miller-Urey experiment) to spontaneously create a self-reproducing ribozyme.

In a ground-breaking experiment, Lincoln and Joyce (Lincoln and Joyce, 2009) used natural selection to help isolate a pair of ribozymes capable of reproducing each other (see Box 6.3). Each ribozyme is a complex molecule made from hundreds of nucleotides, but even so this is one of the first examples of **self-replication** (or perhaps we should call it pair replication) using only ribozymes. In fact, the pair-replication, where two RNA molecules catalyze each other's synthesis, is so efficient that the number of ribozymes grows exponentially (in a controlled laboratory setting). The question remains as to whether such a system (a pair of ribozymes) could have spontaneously

---

[3] See, for example, Zhou, O'Flaherty, and Szostak (2020), where the ribozyme is only fifty-two nucleotides long.

occurred when starting just with organic building blocks, in an environment like the Miller-Urey experiment.

Box 6.3 Self-Replicating Ribozymes

Much progress has been made in showing that ribozymes can replicate under the right conditions. For example, Joyce and collaborators (Joyce, 2009; Robertson 2014) have shown that when a test tube is prepared with two building blocks (each are RNA molecules having many dozens of nucleotides), then these ribozymes can react together, doubling the population within minutes. Of course, this test tube environment is not likely to occur under natural conditions, where long RNA strands (like the building blocks in the test tube) don't seem to occur spontaneously. Still, the mere fact that ribozymes can replicate so quickly, without the presence of any protein enzymes, would have astonished scientists in prior decades.

Rather than just ask whether ribozymes can replicate, the more interesting question is whether they can evolve (from spontaneous mutations) to form RNA molecules that are more efficient at replicating or could copy a template of shorter RNA strands. Such experiments have been done (Joyce, 2009), and the results are very encouraging. The goal here is to show that RNA can, at least under controlled conditions in the lab, undergo the same random mutation process that is the basis for natural selection (as Darwin proposed).

Science often makes progress in small steps, and the experiments by Joyce and others is making progress one step at a time. It would require an entire book to delve into the many steps that have been made toward understanding test tube evolution (see for example, the book by Mesler and Cleaves, *A Brief History of Creation*), but the pace of progress is truly remarkable. The bottom line is that the RNA world hypothesis appears very plausible. Step by step we will likely find simpler starting points for ribozymes to be assembled and to reproduce themselves.

Could life have started from a batch of chemicals? Science keeps chipping away at this problem. It looks more and more like a solution is just on the horizon. We may never know the true answer because we can't go back in time to the early Earth, but having a plausible hypothesis that can be tested by science is the next best thing.

One of the most promising avenues to find an RNA molecule that can reproduce itself is called **test tube evolution**, which has been used extensively by Joyce and collaborators (Joyce, 2009). By starting with a ribozyme based on a RNA molecule found in nature, but modified using modern genetic engineering, it will evolve to become more efficient at copying short RNA strands in a nutrient rich test tube. The more efficient ribozyme (which have slight mutations in its genetic code) can be isolated and further evolved in a second experiment. This process can be repeated, allowing natural selection to find a ribozyme that can copy longer and longer strands of RNA. Recently, Joyce and collaborators (Tjhung, Shokhirev, Horning, and Joyce, 2020) used test tube evolution to find a ribozyme capable of copying its ancestor (the original, unevolved RNA molecule), albeit with a small efficiency, in their controlled laboratory environment.[4] While this isn't the same as finding a ribozyme that can copy itself, it's tremendous progress in this direction.

It's not for lack of trying that scientists haven't been able to achieve the goal of finding out how life on Earth got started. It's an extraordinarily difficult problem. Some scientists have speculated that other molecules (Cairns-Smith, 1966), not the nucleotides used to build RNA, could have formed the first self-reproducing molecule, which was later replaced by a more efficient process. The possibilities are nearly endless. It would take a long time to try out all the ideas, and most microbiologists have focused on more practical problems, such as trying to cure cancer. Until a breakthrough comes, how life on Earth got started will remain a mystery.

6.1 Mesmerizing Membranes

All living cells today have an outer "coat" called a **membrane**. The membrane serves a crucial purpose in keeping the good stuff (the chemical machinery that keeps the cell functioning) inside and the bad stuff out. Here, the bad stuff would be any chemicals that could disrupt cellular functions or that could take over those functions, such as a virus. The interesting thing here is that the membrane is manufactured by the cell's inner functions, and so the question arises as to which came first: the protective membrane (which allows the cell to function) or the cell function (which makes the membrane). Obviously, this is another chicken-and-egg paradox.

Of course, it's possible that a primitive version of a membrane could have occurred spontaneously, independent of ribozymes, and then the two somehow merged symbiotically. So, it's natural to ask what a primitive membrane might look like. To do that, we need to know a bit about the chemistry of a membrane.

---

[4] Other groups have also employed test tube evolution, with varying degrees of success.

Today's cell membranes are based on molecules called **phospholipids**, which have a phosphate group at the end attached to a long fatty-acid (lipid) tail. **Fatty acids** are familiar from ordinary soap and are also found occurring naturally in meteorites (Deamer, 1985). Both ends of the phospholipids are hydrophobic, repelling water, and form into spheres, much like soap bubbles do. The phosphate group prevents the transport of harmful chemicals across the boundary, giving protection to the stuff inside. Without the phosphate group on the end, fatty acids still form tiny spheres in water but are more permeable to chemicals.

Jack Szostak[5] and his colleagues did a brilliant experiment (Budin and Szostak, 2011) where they added a small amount of fatty acid to water, which collected into microscopic spheres, and then slowly added more fatty acid, expecting the spheres to grow larger. Instead, what they found to their surprise was that the spheres transformed into long cylindrical filaments that, after slight agitation, divided up into more microscopic spheres. Had they just seen the precursor of cell division, without any cellular machinery inside? Indeed, it is very tempting to make that connection. They further added phospholipids to the mix and found that these molecules were readily taken up by the spheres. So, it's easy to surmise how the transition to modern cell membranes might have occurred.

It is interesting to speculate whether a ribozyme once existed that could catalyze the transition from a fatty acid to a phospholipid. If so, it would have a selective advantage, perhaps by stabilizing the lipid sphere into a more robust object that repelled other molecules. However, the other side of this coin is that the molecules inside would be sealed off from outside nutrients (Budin and Szostak, 2011). Today's membranes have special units called **transporters**, complex assemblies made from protein, that allow only nutrients to pass across the boundary and keep out harmful molecules (such as viruses). Early membranes would have needed a much simpler chemical pathway to help transport nutrients to the interior.

While much more research needs to be done, we have the first hints of a possible solution to this chicken-and-egg problem. If fatty acids were delivered to the early Earth by meteorites (Budin, Debnath, and Szostak, 2012) and collected in sufficient concentration to form microscopic spheres, then maybe a primitive membrane could form, waiting for the right ribozyme to come along. This seems like a lot of "maybes," but it's the best argument we have at

---

[5] Nobel Prize in Medicine 2009, for an earlier discovery of how chromosomes are protected by telomeres.

present to explain how early cells (with membranes) could have formed. One fact remains: essentially all life today functions inside of a membrane, with water inside.

6.2 Water, Water Everywhere

The importance of water to life (as we know it) cannot be overstated. Chemical reactions proceed more easily in a liquid, where the molecules can move around freely yet are close enough to bump into each other. In a solid, like ice, the molecules are trapped in place, and chemical reactions proceed very, very slowly. In air, molecules can move freely, but they are much more spread out, giving less opportunity for chemical reactions to proceed. In addition, water dissolves many chemicals, such as common salt. (The chemical symbol for salt is NaCl, where Na is for sodium and Cl is for chlorine.) Water breaks the bond between atoms so that other reactions can occur, such as Cl breaking free from Na and attaching itself to some other molecule. Water is one of the most solvent of liquids, and it is also abundant in our solar system.

People have speculated that some different form of life could exist using a different liquid, such as **liquid methane**[6] on the surface of colder planets or moons. So far, no signs of life have been found in these cold environments of our solar system. Others have suggested that life could exist in the dense gas clouds such as on Jupiter, but again space probes have found no evidence for life out there. For now, let's assume that water is essential for life and just consider what's needed (from a cosmic viewpoint) to allow life to form.

For a planet to have liquid water, the temperature needs to be just right. In other words, the planet needs to be just the right distance from a star so that the average temperature and pressure are good for water. Then the planet is said to be in the "Goldilocks zone," where the temperature is not too hot, not too cold, but just right. So, this is the balance point for a planet to be right for life.

When the Earth was forming, our sun was cooler than it is today (based on physical models of star formation). Back then, both Venus and Mars were also in the Goldilocks zone and likely had liquid water. Mars had a thicker atmosphere, which enabled it to retain heat (via the greenhouse effect), but over billions of years most of its atmosphere has been lost to space because of its low gravity. Venus, with its CO_2 atmosphere, got warmer over billions

---

[6] Liquid methane is not as soluble as water for many substances, due to the types of intermolecular interactions that water can form and which methane cannot.

of years as the sun grew hotter. This led to a runaway **greenhouse effect** that evaporated all water on Venus, leaving it hot and barren on the surface.

On Mars, there is now good evidence (see Box 6.4) that water flowed on its surface in the past. This has led scientists to speculate that life could have formed on Mars and then was transferred to Earth. How could life get from Mars to Earth? Well, primitive cells, living inside cracks of Martian rocks, could be blasted into space from large meteorites hitting Mars. Those rocks would wander around in space, with spores inside, and then a lucky few could fall onto Earth, releasing the life inside after they hit. This may seem like a far-fetched scenario, but it's entirely possible. Geologists routinely collect meteorites on Earth that have the distinct mineral composition of Mars. Since it happens today, it also happened when the solar system was new.

Box 6.4 The Mars Rovers

Was there once life on Mars? The only way to know is to go there with some scientific instruments. The most practical way to explore Mars at present is via robotic probes like NASA's rovers. One big advantage of sending the rovers (compared with human exploration) is that they don't need a return trip back from Mars. This reduces the cost by a huge amount (lower by a factor of hundred or more).

The first Mars rover, called *Sojourner*, was small (about the size of a child's tricycle) and weighed only 25 pounds. It had six wheels, designed to go over rocky ground, and a solar panel for power. Sent to Mars in 1997, equipped with just a camera and an X-ray spectrometer, it was really a test of the rover concept and not designed for sophisticated experiments to detect life. But the pictures they sent back caused a worldwide sensation.

The next rovers were bigger, about the size of a small car and weighing about 400 pounds. Two identical rovers, called *Spirit* and *Opportunity*, with more elaborate scientific instruments landed on Mars in 2004 at two different locations. Both rovers explored Mars for over five years. The mission goal was to find out whether water had once existed on the surface of Mars by looking at the geological evidence. While some evidence for water long ago on Mars was found, there was no evidence of microbial life in the past.

The latest generation of car-sized rovers are named *Curiosity* and *Perseverance*. They landed on Mars in 2012 and 2021, respectively, and both are still operating. With more sophisticated instruments, the mission goal is to find evidence of whether Mars could once have supported life. In 2018, organic molecules of benzene and propane were found in three-billion-year-old Martian rock samples by *Curiosity*, but this is not proof of microbial life.

The unique aspect of *Perseverance* is its ability to store cored-rock samples in tubes that could, potentially, be returned to Earth for laboratory analysis. Although the return mission is still being planned by NASA and collaborators, this is the most practical way to get pristine samples of Martian rocks in front of humans. It is much easier to do a robotic reconnaissance mission than sending humans to Mars and back (and less dangerous).

Whether microbial life once existed on Mars is still an open question. The advances in technology have made robotic probes that have been extremely successful in getting close to an answer. Also, the prospect of getting Martian rock into laboratories on Earth seems likely within the next decade. With sophisticated equipment on Earth, the answer may lie within reach.

The bigger question is whether life could have begun on Mars, and if so, whether it also could have independently formed on the primitive Earth. Billions of years ago, liquid water was plentiful on both Earth and Mars. Both planets had nearly the same environment for life to start. If the chemistry of life can start spontaneously, due to random collisions of organic molecules (nucleotides and/or amino acids), then couldn't it have been happened twice, once on Mars and once on Earth? This question has led NASA to make the exploration of Mars one of its primary missions. Today, Mars appears to be barren of life, and the only way to tell whether life was once present there is to examine Martian rocks with sensitive instruments that could find evidence of primitive life from long ago. Sending humans to Mars at this point would be detrimental because that would contaminate Mars with microscopic life from Earth, which would presumably find a way to adapt to the Martian environment.

Suppose that NASA were to find evidence of primitive life on Mars, either past or present. How would we know that it formed independently from Earth? After all, if it's possible that life could have gone from Mars to Earth via rocks blasted into space, then going the opposite direction is also possible. Life from primitive Earth could have already reached Mars eons ago. How would you know whether life on Mars wasn't seeded from primitive life on Earth?

The key to determining whether life started just once, or more than once, is to compare the chemistry of that life. Chemistry might not always take the same path. There are multiple solutions possible to get replicating molecules. Life on Earth may have taken one path initially and then branched out with many different types of reproductive chemistry. Over time, we believe many forms of primitive life took hold. Natural selection led to the more efficient chemistries surviving. That selection process depends, of course, on the environment. The natural selection that occurred on primitive Earth would almost certainly be different from that on Mars.

If found, primitive life on Mars would likely differ in its chemistry than life on Earth. But some things may be similar, such as the method of carrying information via RNA or DNA. If there's some other viable method to chemically transmit the code of reproduction than RNA/DNA, then it would make a worldwide sensation with the announcement of an alternative form of life. There is a whole literature of hypothetical alternatives to the chemistry of life,[7] but no one has yet produced a life form in the laboratory that is vastly different from life as we know it. In any case, it would be very enlightening if any evidence of any primitive life were found on Mars, past or present, and whether its chemistry would be different from life on Earth.

6.3 Right-Handed and Left-Handed

Rather than exploring all the possible alternatives that have been proposed for the chemistries of primitive life, let's focus on a particular case, where a slight imbalance could tip the chemistry of life in one direction or another. Consider the case of some molecules that can come in two variations, called left- and right-handed.

Comparing your left and right hands, you see a **mirror symmetry**. If someone shows you a picture of a hand, it's easy to tell whether it's a right or left hand. In scientific terminology, the hand is a **chiral** object, meaning that it has a distinct left or right shape, with mirror symmetry. Take that same picture and look at it in a mirror, and it appears to have opposite chirality: a picture of a left hand looks in the mirror like a right hand.

Some simple molecules, such as sugar, are chiral objects, and can be made in the laboratory with either a left- or right-handedness. Just like a hand, the molecule has a "front" and a "back," which depends on the positions in space (see Figure 6.2) of the atoms that make up the molecule. Look again at your hands. If you rotate your right hand by 180°, it has the same outline as your left hand, but of course you can tell the difference by observing whether the palm is facing toward you or not. The same principle can be used to define the chirality of a molecule.

---

[7] See, for example, A. F. Davila and C. P. McKay (2014) *Astrobiology*, 14, pp. 534–540.

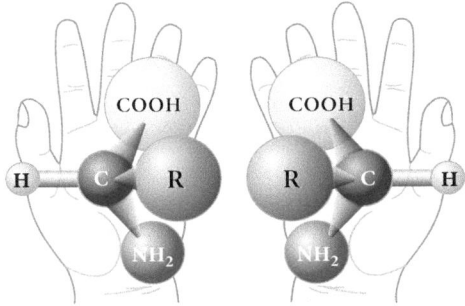

Figure 6.2 The handedness of a generic amino acid molecule, where R labels the "side chain" of the amino acid. In the laboratory, both left and right chirality can be synthesized, as shown. The human hands in the background are for illustration only.
Reproduced from Perhelion (2011). Via Wikimedia Commons (public domain).

In the laboratory, for setups such as the Miller-Urey experiment, both left-handed and right-handed molecules of simple building blocks, like sugars and amino acids, are produced with equal probability. But all sugars used by life on Earth are right-handed (see Box 6.5), and almost all amino acids are left-handed.[8] This is true for the simplest organisms like single-cell bacteria and for more complex organisms like mammals. Similarly, all DNA and RNA is made from only right-handed molecules. Why life evolved to use just a single chirality of these molecules is a mystery.

Box 6.5 Louis Pasteur and Polarized Light

In nineteenth-century France, the topic of spontaneous generation was all the rage. One side argued that the fossil records showed that different species had many similarities and that all vertebrates had likely come from a common ances-tor, thus suggesting that life appeared spontaneously and then evolved. The other side was firmly on the side of the Catholic Church, believing that all creatures were created by God. To settle this issue, debate was initiated on this topic by the French Academy of Sciences, with the winner to receive the Alhumbert Prize and 2,500 francs. The debate was between naturalist Felix Pouchet and chemist Louis Pasteur.

Pasteur was born in 1822 into a poor Catholic family. At school, he was an average student, with more interested in sketching and painting than science. At college, he

continued

---

[8] There are some biologically important D-amino acids (right-handed), but they are rare.

Box 6.5 *continued*

---

earned a degree in mathematics but struggled with chemistry. In 1843, he entered the prestigious *Ecole Normale Superieure*, eventually submitting two theses, one in chemistry and one in physics. His studies in physics involved polarized light, and in particular how polarized light could be rotated when passing through certain liquids. Later, as a Professor of Chemistry at the University of Strasbourg, he would put his knowledge of physics to good use.

At Strasbourg, Pasteur encountered a perplexing result regarding tartaric acid, which is found in fruits such as grapes. When polarized light passes through this liquid, it rotates the polarization clockwise. There was also a synthetic version of tartaric acid, sold to bakers under the name racemic acid, which didn't rotate the polarization. As a chemist, Pasteur knew the chemical compositions were identical. Why would tartaric acid rotate polarized light but racemic acid would not? He had discovered that life only produces one chirality of the molecule, whereas synthetic chemistry produces both left- and right-handed molecules equally. Although the significance of this was not known at the time, it would later be shown that all known life uses one chirality, suggesting that all life comes from a common ancestor.

The irony of this story is that Louis Pasteur won the Alhumbert Prize for his argument that spontaneous generation didn't occur. He did this based on careful experiments showing that substances such as milk, wine, and beer could be sterilized by boiling, a process now called pasteurization. Using a swan neck flask, which prevented dust from settling on the sterilized liquid, and pinching off the end using glass-blowing techniques, he showed that the contents would not spontaneously generate life. At the debate, Pasteur stated that those believing in spontaneous generation must conclude that "God is useless."

While Pasteur's accomplishments in medicine and chemistry are certainly his greatest contributions to society, his discovery that life uses chiral molecules should not be forgotten. The possibility that alien life having molecules with the opposite chirality might exist is very tantalizing!

One can imagine a mirror-world scenario where extraterrestrial life would use the opposite molecules, left-handed sugars and right-handed amino acids. Organic chemistry would work just as well with the mirror-image of all molecules. Obviously, finding evidence in Martian rocks of organic compounds having the opposite chirality would be a huge discovery, suggesting that life started independently there. Sadly, no complex organic compounds of either left- or right-handed have yet been found on Mars.

There must be some balance point that tipped life on Earth in the direction of using one chirality of molecules. In a classic biochemistry experiment, Jerry Joyce (Joyce et al., 1984) measured the synthesis of RNA (using dedicated chemical reactions, where enzymes were added to the mix) in a soup of molecules where the building blocks (nucleotides and sugars) were purely of one chiral type. But when the soup contained both chiral types of molecules, formation of the RNA strand was greatly inhibited. This suggests that for RNA (or a ribozyme) to occur naturally, nature needed a way to purify the soup, so that only one chiral type remained. Some imbalance had to occur that tipped the chemistry of life toward using just one-handedness of molecules.

Joyce and his group continued to experiment with left- and right-handed nucleotides and eventually found a symbiotic pair of ribozymes where the left-handed ribozyme (or L-RNA) could catalyze the assembly of its mirror-image right-handed ribozyme (or D-RNA) and vice versa (Sczepanski and Joyce, 2014). Both ribozymes had just eighty-three nucleotides. The process is remnant of the well-known drawing by Escher (see Figure 6.3) where the left hand is drawing the right and vice versa.

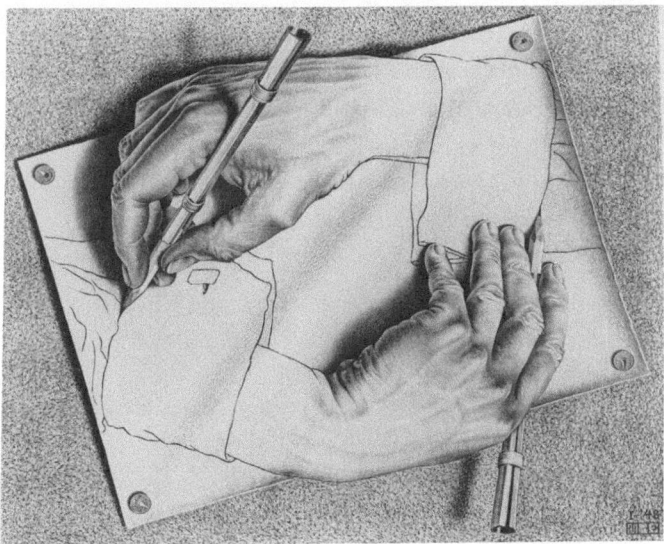

Figure 6.3 The work of artist M. C. Escher titled "Drawing Hands" from 1948. It can be likened to the experiment by Sczepanski and Joyce where D-RNA and R-RNA catalyze the replication of each other.

The **cross-chiral** experiment described above is like the pair-replication experiment described earlier (which used only D-RNA) but now with mirror RNA molecules. It suggests the possibility of competition in the prebiotic soup between L-RNA and D-RNA. For whatever reason, the D-RNA won out. Once the D-RNA could replicate faster, it would take over by natural selection. These experiments, as ingenious as they are, still only hint at what might have happened on the early Earth. Indeed, alien life might have evolved in a way that incorporates molecules of both chirality.

Since we don't know how life started from a soup of building-block molecules, it seems pointless to speculate on whether other life forms, either on Mars or elsewhere in space, would use the same chemistry, or even the same chirality of molecules as used by life on Earth. It seems that there must be balance points along the way if life developed spontaneously from naturally occurring molecules. This is perhaps the next great challenge to science, to learn how life on Earth started, or how life elsewhere could be different. It is probably a greater challenge than figuring out how the universe started.

Physical laws, such as quantum physics and general relativity, can be described by a set of equations that obey specific rules (or "laws"). However, the chemistry of life has a complexity that can't be boiled down to a set of unique equations. Exploring all the possibilities of how a soup of molecules turns into a set of self-replicating ribozymes (or some other alternative) could take several lifetimes, even with a massive worldwide effort. However, scientists are clever, and perhaps a major breakthrough is just a few years away. One thing is clear: this is a rich field, ripe for discoveries. Learning about the origin of life will add to the already impressive list of the balance points that exist in Nature to make life on Earth possible.

References

Budin, I., and Szostak, J. W. (2011) *PNAS Research Article*. Available at: https://doi/10.1073/pnas.1100498108.

Budin, I., Debnath, A., and Szostak, J. W. (2012) "Concentration-Driven Growth of Model Protocell Membranes," *J. Am. Chem. Soc.*, 134, pp. 20812–20819.

Cairns-Smith, A. G. (1966) "The origin of life and the nature of the primitive gene," *Journal of Theoretical Biology*, 10, pp. 53–88.

Zaug, A. J. and Cech, T. R. (1986) "The intervening sequence RNA of Tetrahymena is an enzyme," *Science*, 231, pp. 470–475. See also: Cech, T. R., Zaug, A. J., and Grabowski, P. J. (1981), *Cell*, 27, pp. 487–496.

Crick, F. (1968) "The origin of the genetic code," *J. Mol. Biol.*, 38, pp. 367–379.

Deamer, D. W. (1985) *Nature*, 317, pp. 792–794.

Guerrier-Takada, C. and Altman, S. (1986) "M1 RNA with large terminal deletions retains its catalytic activity," *Cell*, 45, pp. 177–183.

Johnston, W. K. et al. (2001) "RNA-Catalyzed RNA Polymerization: Accurate and General RNA-Templated Primer Extension," *Science*, 292, pp. 1319–1325.

Joyce, G. F. et al. (1984) "Chiral selection in poly(C)-directed synthesis of oligo(G)," *Nature*, 310, pp. 602–604.

Joyce, G. F. (2009) "Evolution in an RNA World," *Cold Spring Harbor Symp. Quant. Biol.*, 74, pp. 17–23.

Lincoln, T. A., and Joyce, G. F. (2009) "Self-sustained Replication of an RNA Enzyme," *Science*, 323, pp. 1229–1232.

Robertson, M. P., and Joyce, G. F. (2014) "Highly Efficient Self-Replicating RNA Enzymes," *Chemistry & Biology*, 21, pp. 238–245.

Sczepanski, J. T., and Joyce, G. F. (2014) "A Cross-chiral RNA Polymerase Ribozyme," *Nature*, 515, pp. 440–442.

Tjhung, K. F., Shokhirev, M. N., Horning, D. P., and Joyce, G. F. (2020) "An RNA polymerase ribozyme that synthesizes its own ancestor," *PNAS*, 117, pp. 2906–2913.

Zhou, L., O'Flaherty, D. K., and Szostak, J. W. (2020) "Assembly of a Ribozyme Ligase from Short Oligomers by Nonenzymatic Ligation," *J. Am. Chem. Soc.*, 142, pp. 15961–15965.

Chapter 7
It Came from Outer Space

If the dinosaurs had had a space program, they would not be extinct.
—**Carl Sagan**

In many ways, it seems like Earth is just too perfect for life. We live in the "habitable zone" of the solar system, where it's not too hot (water doesn't boil) yet not too cold (water remains liquid). Astronomers call it the **Goldilocks zone**. Aside from its placement in the solar system, there are other ways in which the Earth is situated just right, such as its placement within the galaxy or its placement in time (measured from the Big Bang). First, let's look at various extra-solar factors that keep life in balance.

There are dangers to life on Earth that come from outer space that aren't obvious. Indeed, most normal people (I'm not counting astronomers!) tend to think of outer space as unchanging, with the stars in the same place every night, immutable. Yet if we look closer, using the Hubble Space Telescope, we find stars that have exploded, like the one shown in Figure 7.1. The Crab Nebula is the result of a supernova that exploded in year 1054 and was recorded by Chinese astronomers. It was about 6,500 light-years away. But what if something like this happened to a star closer to Earth? Could an explosion in space light-years away us wipe out all life on Earth? How precarious is our existence, on the cosmic timescale?

In this chapter, we'll look at a few examples of how life on Earth is lucky to have survived. It's like we are winning a cosmic roulette game, where the next spin could spell disaster. Whether it's meteorite impacts from the sky or gamma radiation from space, our lucky streak continues. But for how long?

7.1 Killer Supernovae

It is no longer fantasy to say that a supernova could have wiped out all life on Earth. We know the rate of supernova explosions and how close a supernova needs to be for the effects to be devastating to life on Earth (see Box 7.1). Earth's sediments have geological proof that dust from a nearby supernova explosion reached us. The idea that this could happen was suggested years

Nature's Balancing Act. Ken Hicks, Oxford University Press. © Oxford University Press (2025).
DOI: 10.1093/9780197771471.003.0007

Figure 7.1 Hubble Space Telescope picture of the Crab Nebula, showing the result of a supernova explosion.
Reproduced from NASA, ESA, Hester and Loll (2005). Via Wikimedia Commons (public domain).

ago by Ellis and Schramm (Ellis and Schramm, 1996). The first evidence was published in 2004 by a German-Austrian team (Knie et al., 2004), who used a device called an atomic mass spectrometer (AMS), which separates isotopes atom by atom. Figure 7.2 shows the results, where a spike in the abundance of a radioactive isotope, **iron-60**, is seen in ocean sediments at about 2.8 million years ago. That isotope is expected to be produced in prodigious amounts by a supernova.

Box 7.1 Supernova Explosions

A supernova is one of nature's most violent events. When a star runs out of nuclear fuel, it can no longer generate the heat and outward pressure needed to oppose the inward gravitational force. The result is that the star collapses, and its gravitational energy is concentrated into a small radius. The star's matter can only be compressed so far before the atomic nuclei bounce off each other, directing the energy outward in a fantastic explosion.

continued

Box 7.1 *continued*

There are several different types of supernovae, but the most common ones are called type I and type II. Type II are easier to explain, as they come from a very massive star, having about ten times more mass than our sun, where the core collapses. When the matter bounces, the outer layers shoot outward, and the inner core may further collapse into either a black hole (Box 7.2) or a neutron star (Box 7.3). The ejected material undergoes rapid nuclear reactions, releasing tremendous amounts of light and radiation over a few days. The light output from a single supernova can outshine the luminosity of an entire galaxy, making them visible to telescopes here.

Type II supernovae are thought to come from binary systems of stars, where two stars rotate around each other. One star, with mass that's only a few times the mass of our sun (so it doesn't become a type II supernova), would slowly cool into an object called a white dwarf star. The white dwarf is just hot compressed matter that isn't massive enough to further collapse into a black hole. If the mass of a white dwarf exceeds about 1.4 times the mass of the sun, then it collapses, causing a type II supernova. To gain mass, the white dwarf needs to steal it from somewhere. The outermost layer of its companion star in the binary system is the likely source, pulled in by the strong gravitational pull of the white dwarf.

Of course, the explanations above are simplified explanations. The full range of supernova observations span a wide range of phenomena. But type II are common enough that they can be used as a "standard candle,"[1] since the light output is nearly the same (coming from nearly the same starting point, with 1.4 solar masses). This enables astronomers to estimate the distance of a far-off galaxy if a type II supernova is seen there. Using this distance, along with the velocity measured for that galaxy, cosmologists can calculate whether the universe is contracting or expanding.

Current research in nuclear physics is helping to understand the process of supernovae explosions. As the outer shell of matter is flung outward with high speed, it undergoes numerous nuclear reactions. Those reactions can be measured at accelerator labs and put into computer models of supernova explosions. The field of nuclear astrophysics is rapidly developing, in part due to advances in accelerator technology.

---

[1] This term is used by astronomers to mean a calibrated light source, giving a known luminosity.

After the iron-60 is ejected from the supernova, it must cross vast regions of interstellar space to reach Earth. How much time this takes can be estimated based on physical models of supernova explosions. From this, along with the iron-60 abundance, scientists can figure out how long ago (and how far away) the supernova occurred.

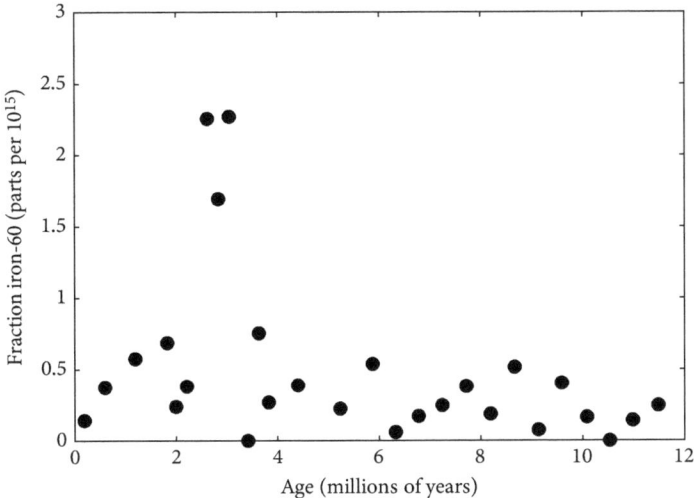

Figure 7.2 The fraction of iron-60 out of all iron content in ocean sediments, versus age of the sediment layer. (Zero is today.)
Adapted with permission from Knie et al. (2004).

Several recent scientific publications (Wallner et al., 2016) have confirmed this unusual abundance of iron-60 in the sediment records of other oceans, and additional evidence comes from higher-than-expected amounts of iron-60 in the lunar soil samples from the Apollo moon missions. The amount found in sediments is much larger than expected if it were due to other sources, such as cosmic-ray interactions with elements in the Earth's crust. This is strong evidence that at least one supernova exploded near Earth during the past few million years.

Iron-60 has a half-life of 2.6 million years, meaning that half of the remaining iron-60 atoms decay to a stable (nonradioactive) isotope every 2.6 million years. For example, at 5.2 million years after a supernova explosion, only one-quarter of the original iron-60 atoms remain, since two half-life periods have passed ($1/2 \times 1/2 = 1/4$). So, the timescale where iron-60 is useful as a marker of supernova explosions is limited to about the past ten million years. After that, most of the iron-60 has decayed away, leaving nonradioactive isotopes.

The rate of supernovae in our Milky Way galaxy is estimated to be about one to two per century. The chances of a supernova going off close to Earth in any given century is very small. But over a timespan of ten million years, the probability of a supernova within about 300 light-years of Earth is significant. The iron-60 evidence shows that this has happened at least once.

To be dangerous to ancient life, a supernova would need to occur at a distance closer than about thirty light-years. In that case, the high-energy radiation from the supernova would ionize atoms in the atmosphere, destroying the protective ozone layer. Without the ozone layer, the harmful ultraviolet (UV) radiation from the sun could reach the ground in force, breaking apart the molecules of life, damaging DNA. In addition, chemical reactions would occur in the atmosphere, forming compounds such as nitrous oxide, a pollutant that causes acid rain.

The probability of a **near-Earth supernova** to occur is small over a period of ten million years. However, recall that the age of our solar system is about 4,600 million years. It is estimated that there's at least a 50% chance that this happened once since the Earth was formed (Gehrels et al., 2003). If it occurred when microbial life had a foothold in the deep ocean, then that life would likely be protected from changes in Earth's atmosphere. However, larger life forms might not be so lucky. One study (Melott et al., 2004) suggests that a gamma-ray burst (see below) is responsible for one of the mass-extinction events that resulted in wiping out nearly 60% of species in the ocean.

Microbial life would also face the danger of neutrinos emitted from a close-by supernova. Neutrinos are given off in huge quantities from supernova, and these particles pass through ordinary matter with ease. Although the biological effect of neutrino interactions is still uncertain, estimates suggest that a near-Earth supernova (Collar, 1996) would cause enough neutrino interactions[2] to damage a significant fraction of life forms bigger than a small fish.

The only supernova in the vicinity of our galaxy that's been seen with modern telescopes occurred in 1987, called **supernova 1987A**. It exploded in the Large Magellanic Cloud, about 168,000 light-years distant. In addition to the light seen from this supernova, neutrino detectors on Earth saw about a dozen neutrino interactions (Arnett, 1989) in a large water detector. Considering that the neutrinos stream out into space in all directions, in this case spreading out over a sphere with radius 168,000 light-years, only a very tiny fraction hit the entire Earth. An even smaller fraction hit the neutrino water detector.

You can imagine the large number of neutrino interactions that would occur from the concentrated blast of a near-Earth supernova. If there are enough neutrino interactions, this would cause damage to the molecules in all living cells, possibly causing a mass extinction of life on Earth. It would almost

---

[2] Estimated at 10,000 neutrino interactions per kilogram of living tissue for a near-Earth supernova.

certainly wipe out all large life forms. Luckily, there's no candidate star to go supernova within thirty million light-years of Earth at present.

One star that is likely to go supernova within thousands of years is Betelgeuse, a red giant star located in the constellation Orion. It is one of the brightest stars in the sky and is easily seen with the naked eye. Its distance is 430 light-years away, which is far enough away not to threaten to life here. Although Betelgeuse might not go supernova until 100,000 years from now (Dolan et al., 2017), the timeframe is uncertain, and some people have suggested that it could happen anytime (even tomorrow). Although the nuclear reactions leading to a supernova are reasonably well-modeled in computer simulations, the observational clues of a star just about to go supernova are purely theoretical. When it does go, the supernova from Betelgeuse will shine as bright as the moon and be visible even during the daytime. In addition, the neutrinos detected from it will provide rich data, giving clues about how these stars explode. It will not, however, pose a danger to us.

7.2 Gamma-Ray Bursts

Supernovae are not the only danger to life on Earth from space. There are other mysterious objects out there that produce vast amounts of gamma rays. To understand what these objects are, let's see how they were detected, and then ask whether they could be a danger to life on Earth.

At the height of the Cold War, the United States put military satellites in orbit so as to detect radiation from atomic bomb tests of other countries such as the Soviet Union. Indeed, these satellites did detect bursts of gamma rays, but surprisingly the bursts were not coming up from the ground but down from outer space. These events were called gamma-ray bursts (GRBs), and the mysterious source was left to the astronomers to figure out.

At first, it was thought that the GRBs were coming from within our Milky Way galaxy. This made sense, because if GRBs were coming from other galaxies, the amount of energy at the GRB source would be immense, more than a supernova. The burst typically lasts only for a few seconds, so all of that energy is concentrated in time. If the burst were going out uniformly in all directions from a source outside of the Milky Way, it would be like the entire mass of the sun turned into pure energy in the span of a few seconds. This seemed so unlikely that it made sense that GRBs were coming from inside our galaxy. It was later learned that this assumption was wrong.

New satellites were put into orbit, made specifically for detecting the locations of GRBs. It didn't take long to realize that the GRBs were coming from all over the sky, and not confined to the plane of the Milky Way. It was also found that GRBs come in two types: short GRBs (lasting less than two seconds) and long GRBs (lasting about ten seconds). It is now believed that there are two different astronomical phenomena that produce GRBs.

A possible source of GRBs is from very massive stars, having twenty-five times the mass of our sun or more, which can collapse their cores into black holes (see Box 7.2) after going supernova. The star's outer layers would then be whipped around the black hole because of the star's rotation. The huge gravitational and electromagnetic fields there create a vortex that funnels matter at near light speed. The star remnants stream outward from its poles in a tight "jet," as shown in Figure 7.3.

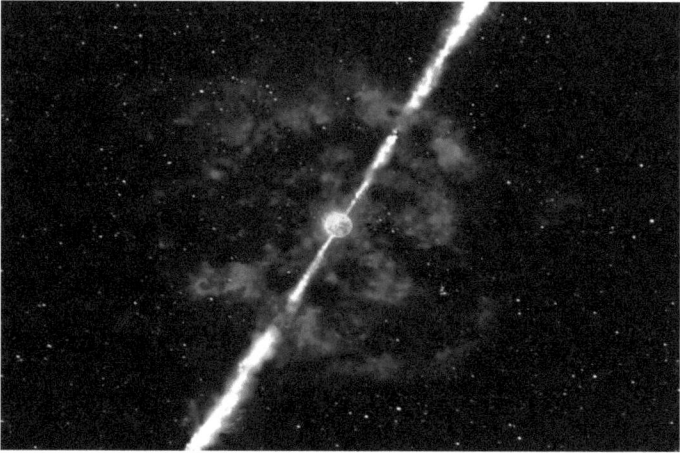

Figure 7.3 Artist illustration a gamma-ray burst (GRB) following a supernova explosion. If the narrow jet is oriented in the direction of Earth, a few-second burst of radiation is detected.
Reproduced from NASA, Zhang and Woosley (2008). Via Wikimedia Commons (public domain).

Box 7.2 Black Holes

The evidence for black holes dates to the early 1970s, from observations of a point source of X-rays called Cygnus X-1, located in the direction of the constellation Cygnus the Swan. However, the first hints that they could exist go back to 1916, shortly after Einstein's theory of general relativity was published. Einstein's equations could be solved giving a gravitational field of a point mass (also called a singularity). For a large enough mass, the gravity would be so strong near it that even light could not escape.

It wasn't until the late 1930s that Oppenheimer and Snyder convinced the physics community that black holes could exist. Using general relativity, they showed that black holes could arise from continued gravitational contraction of matter. The term "black hole" was coined by physicist Robert Dicke in the 1960s and later popularized by the eminent physicist John Wheeler.

Unlike stars, which come in all sorts, from red giants to white dwarfs, black holes are featureless. Because black holes have no visible characteristics, it is common to hear the phrase "black holes have no hair." The no-hair theorem, as it's called, states that only the mass, charge, and angular momentum (or spinning) give a black hole its distinguishing characteristics. In other words, all black holes look the same. Presumably any cosmetic feature ("hair" being the analogy) would contradict the idea that nothing (not even light) can escape from a black hole.

Most black holes can be categorized loosely into two types: massive black holes having tens to hundreds of solar masses, and supermassive black holes having millions (or billions) of solar masses. The former are thought to be the remnants of supernova explosions, and the latter seem to be at the center of every galaxy (including our Milky Way). Only recently have intermediate black holes with thousands of solar masses been found in smaller galaxies,[3] but these are rare. How the supermassive black holes formed is still a mystery. This remains a hot topic of astronomy research.

There is another aspect where black holes are unique. They attract matter and antimatter equally. One could imagine a situation where one black hole is made mostly from matter and another is made mostly from antimatter, and because of the no-hair theorem, the two would appear the same.[4] Could black holes have something to do with the matter dominance of our universe? No one has suggested that this could explain the matter-antimatter imbalance of the universe (Chapter 3). But the mystery remains of how supermassive black holes formed, and the role of black holes in the early universe is largely unknown.

[3] Evidence is from the galaxy Omega Centauri, see M. Haberle et al. (2024), *Nature*, 631, pp. 285–288.
[4] E. Witten, American Physical Society April Meeting 2016, session W1, "Symmetries and Geometry."

It is now well established that jets of material are ejected from the center of many galaxies, where supermassive black holes are located. Observations from X-ray telescopes like the orbiting CHANDRA satellite have shown clear pictures of these jets. It's easy to surmise that a black hole created by the death of a very massive star would have a similar jet, lasting for tens of seconds. This is believed to be the origin of the so-called **long-GRBs**. We only see those bursts when the jet happens to be pointed almost directly at Earth. Because of the tight beam of gamma rays, the amount of energy released is much smaller than if the same intensity of gamma rays were spread out uniformly in all directions. Doing the math, one finds that the energy release is consistent with theoretical models of a jet following a supernova.

The origin of the short-GRBs is thought to be different. While an afterglow in visible light was seen correlated with some of the long-GRBs, consistent with the belief that they are associated with a certain type of supernova, no afterglow was seen for the short-GRBs until recently (from an event that was unusually close). The current best guess is that short-GRBs come from the merging of two neutron stars. This happens when neutron stars (the remnants of large dying stars, see Box 7.3) come in binary systems, orbiting a common center of mass. When the neutron stars lose energy (due to emitting gravitational waves), their orbits shrink, and the two stars eventually merge in a cataclysmic event, likely resulting in a black hole that emits a short burst of gamma rays. The hypothesis that GRBs are due to neutron star mergers was strengthened when a short, small burst of gamma rays was seen at the same time as gravitational waves were detected by LIGO (Abbott et al., 2017). This event, called GR170817, was close enough (in a galaxy not so far, far away) that telescopes could catch the optical afterglow. Now called a kilonova, meaning as bright as 1,000 novae, these events are a possible source of short-GRBs.

Box 7.3 Neutron Stars

Neutron stars are one of the weirdest objects found in outer space. A teaspoon full of the matter on its surface weighs more than an entire mountain on Earth! This is because the matter has been so compressed by gravity that it's like a giant nucleus, packed into a radius of about 10 km (the size of a small city).

In normal matter, the nucleus is much smaller than an atom (which is smaller than any optical microscope can see). The atom is mostly empty space, with electrons orbiting the nucleus. If you could zoom in to see it, the electron orbits are about 10,000 times bigger than the nucleus. To compress matter, you need to get rid of the electrons, which repel each other (keeping atoms apart). This can be done, at the cost of enormous energy, by absorbing the electrons onto the protons, via the weak nuclear process. In a neutron star, that energy comes from gravity, such as in a core-collapse supernova (Box 7.1).

The bottom line is that neutron stars can't be made from normal interactions of matter. It takes the power of a supernova, which leaves a neutron star (or a black hole if the progenitor star is large enough). So how do astronomers observe neutron stars? It turns out that neutron stars rotate, and as they do so, they send beams of radiation out from the poles. When that radiation is pointed toward Earth, radio telescopes hear a distinct blip each time it goes around. This regular pulsation was detected long ago and called pulsars by astronomers before it was known that neutron stars were the cause.

Estimates[5] show that a billion neutron stars exist in our Milky Way galaxy. One of the most well-known is at the center of the Crab nebula (Figure 7.1), located in the constellation of Cancer the Crab. The Crab pulsar emits radiation from low-energy radio waves up to high-energy X-rays. The large number of neutron stars is due to the age of our galaxy, which has had billions of supernovae over its lifetime.

Some of neutron stars even come in binary systems, with two neutron stars orbiting their center of mass. These systems will radiate gravitational waves, causing them to lose energy and their orbits to shrink, as first measured by Hulse and Taylor.[6] Over time, they will merge in a cataclysmic event (see text).

Neutron stars are the subject of intense study by theoretical nuclear physicists because they tell us about the properties of compressed nuclear matter. Theorists speculate that the center of the neutron star, where the pressure is highest, could have stable amounts of strange quarks mixed in with the neutrons, called strange matter. Or it's possible that quarks could become free from the confines of neutrons, becoming quark matter at the center. There's a lot of new knowledge to be gained from neutron stars.

[5] https://imagine.gsfc.nasa.gov/science/objects/neutron_stars1.html
[6] Russell Hulse and Joseph Taylor, Nobel Prize in Physics 1993.

So, what's the danger to Earth? If a GRB were close enough to Earth, with the jet directed toward us, the gamma rays would destroy the ozone layer and, for land-dwelling mammals, let in enough ionizing radiation that cancer or death would result. Like supernovae, the rate and distribution of GRBs is known. One paper (Piran and Jimenez, 2014) has calculated that there is a 50% chance that a lethal GRB event took place on Earth during the past 500 million years. The rate of GRBs increases with the density of stars, such as near the center of our galaxy. Also, GRBs were more plentiful in the early universe. Piran and Jimenez conclude that the safest environments for life (as we know it) are on the outskirts of the galaxy, much like where Earth is located (see the conceptual picture in Figure 7.4). The authors conclude that for the inner part of our galaxy, about 12,000 light-years from the center, there is a 95% chance of at least one lethal GRB over one billion years. Such a scenario would be very inhospitable to life. So, Earth seems to be in a lucky location.

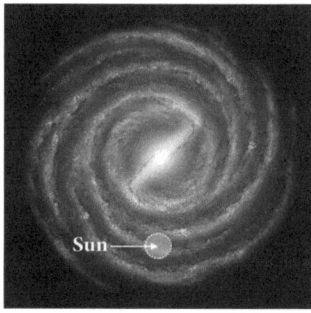

Figure 7.4 Artist illustration of the Milky Way Galaxy showing the location of our sun.
Reproduced from NASA/JPL-Caltech, Pyle (2021). Via Wikimedia Commons (public domain).

7.3 What Killed the Dinosaurs

A more immediate danger to human life is the possibility of a medium-sized asteroid or comet to cross the path of Earth's orbit and smack into the ocean, like what happened sixty-six million years ago when the dinosaurs went extinct (at the **K/T boundary** in the fossil record). Although near-Earth supernovas or GRBs pointed toward us are rare events, meteorites striking the Earth are quite common, at least for small-sized objects. Large objects capable of destruction on the scale of a thousand atomic bombs (20 Megatons of TNT equivalent) happen on average once every few centuries. Titanic events, like the one that led to extinction of the dinosaurs (see Box 7.4), happen once every hundred million years. The relation between object size and time between impacts is shown in Figure 7.5.

Box 7.4 What Killed the Dinosaurs?

In the fossil record, there is a clear dark line in the sediment layers called the K/T boundary (between the Cretaceous and Tertiary periods). Below the K/T line, you can find dinosaur fossils and not above it. This suggests a sudden disappearance of the dinosaurs at about sixty-six million years ago. What could have caused this dominant species to go extinct?

A clue is in the chemical composition of the K/T boundary. Those sediments are rich in the element iridium, hundreds of times more than in the layers above and below it. Where did the iridium come from? Meteorites have high levels of this rare element, and this led physicist Luis Alvarez[7] and his geologist son, Walter Alvarez, to propose that a huge impact event could have caused this mass extinction (Alvarez,

Alvarez, Asaro, and Michel, 1980). This became known as the Alvarez hypothesis and was met with skepticism at first because there was no impact crater known at that time. Still, the fact that unusually large iridium concentration was found at the K/T line in sediments all over the world was compelling.

Fast-forward, and a huge impact crater was found near the Yucatan Peninsula in Mexico. The reason it had been missed is partly because most of it is under the ocean and partly because it was so large. There is no crater bowl to be seen from the air, with a diameter of over ninety miles. It was discovered by geophysicists looking for underground oil reserves in that region. Estimated to have formed at about the time of the K/T boundary, this was the "smoking gun" evidence that has led to widespread acceptance of the Alvarez hypothesis.

Of course, the dinosaurs never knew this impact was coming. Until recently, humanity would have been unaware of an impending large impact event. It was only in 1998 that NASA set a goal of finding 90% of all **near-Earth objects** (NEOs) with diameter greater than 1 km. Compare this with the K/T impact, which came from a NEO of diameter 10–15 km.

It was also in 1998 that the movie *Armageddon* was released. It's probably not pure coincidence that NASA set its goal in the same year. In 2005, Congress upped the goal to find 90% of all NEOs with diameter greater than 140 meters. One facility that was constructed for this purpose is Pan-STARRS, located in Hawaii, which can map 1,000 square degrees of the sky each night. It has been operating since about 2014.

Even with these telescopes guarding us from NEOs, they aren't perfect. There can be large uncertainties to the obits of asteroids. Based on risk estimates, we appear to be safe from a large impact in the next decades. But if a large comet comes at us from the other side of the sun, we might get little advance notice. That said, the risks appear quite low at present, and the technology to detect a large impact risk is getting better every year.

---

[7] Nobel Prize in Physics 1968, for discoveries in particle physics.

The straight-line fit shown in Figure 7.5 is based on a lot of information, including data from craters on Earth (such as Meteor Crater in Arizona) and the record of craters on the moon (which didn't hit Earth but help to determine the rate of such impacts). Note that the graph in Figure 7.5 has its axes labeled with powers of ten, showing that the straight line obeys a "power law" form, a well-established mathematical relationship for chaotic phenomena. Many books and websites describe in detail the data used to derive this relationship between meteorite size and time between impact, but the important point is that scientists know the risk factors for such impacts.

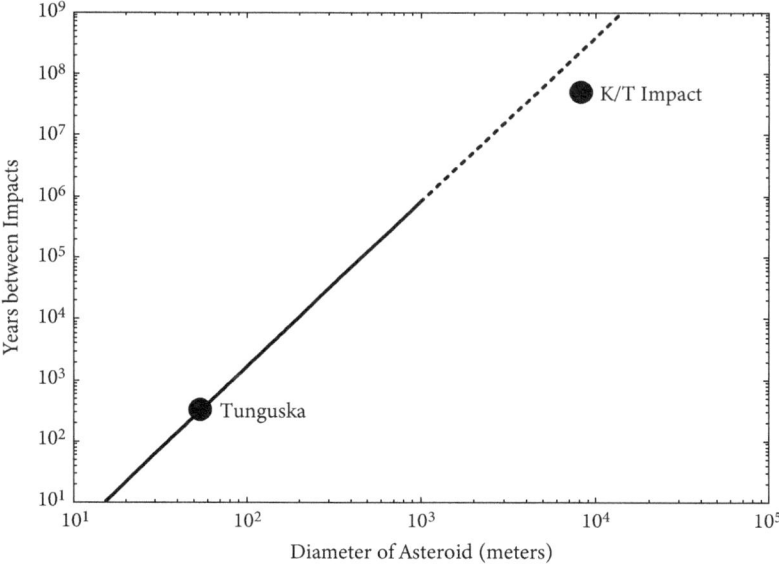

Figure 7.5 Recurrence rate of impacts for different-sized meteorites. Catastrophic impacts from asteroids with diameters greater than 1 km occur about every million years. The solid line is fit to data, the dotted is projected.
Adapted with permission from Rampino and Haggerty (1996).

Of course, a risk factor doesn't tell you anything about what might happen tomorrow. Risk factors are averaged over a large number of events. For example, I might have a low risk factor for cancer, but I could be diagnosed with cancer tomorrow. Even top athletes with excellent diets and low hereditary risks get cancer. Similarly, just because the risk factor for a catastrophic meteorite is estimated at once every hundred million years, it's still possible to happen sometime in the next hundred years. For this reason, several countries around the world have undertaken sky surveys looking for possible **near-Earth objects** (NEOs), with potential impact paths.

A rich source of asteroids is from the asteroid belt, where rocks of varying size orbit the sun just outside the orbit of Mars and inside the orbit of Jupiter. The asteroid belt is a chaotic system, meaning that slight changes to the orbit of one asteroid can influence several other asteroids, cumulating in a large effect on the orbit of another unfortunate asteroid that gets pulled out of its orbit. These somewhat random interactions happen continuously. Sometimes an asteroid is deflected into a path that could strike the Earth. It's almost impossible to predict when this might happen. But with the NEO sky surveys, we can catch any potentially hazardous asteroid well in advance of it hitting Earth. Comets, which originate from beyond the orbit of Pluto, are also tracked, but they have a much lower risk of Earth impact.

7.4 Winning the Lottery

Although the risks are low at present for life on Earth to be wiped out by either a meteorite impact, a GRB, or a nearby supernova, over the timescale of about 3.8 billion years (when microbial life first appeared on Earth) many life-threatening events must have occurred. You might even say that it's surprising that life on Earth has survived so long! And for life on an Earth-like planet closer to the center of the Milky Way galaxy, where the density of stars is higher, the rate of GRBs and other life-threatening events from outer space could very possibly have wiped out any microbial life on planets there.

Is Earth the Goldilocks planet, as some have called it? Are the conditions just right for life, and is Earth just lucky enough to have escaped the equivalent of being "eaten by a bear"? One thing is clear. We haven't encountered evidence of life elsewhere in the galaxy, at least not yet. Even though the balance of Nature's physical laws have made conditions ripe for life to exist, allowing a variety of complex animals to form, the dangers to life continuing (at least long enough to form intelligent beings) on other planets are substantial.

In the past, whenever people have assumed that Earth is in a special place, science has upset that apple cart. Far in the past, most people believed that the sun orbited the Earth, until Copernicus showed that Earth was not the center of the solar system. Similarly, when it was assumed that all stars resided in the Milky Way, it was discovered that some stars that appeared "fuzzy" in the telescope were actually far-off galaxies and that our Milky Way was not a special place in the universe. Before about 1995, it was not known whether planets orbited other stars, but today astronomers have shown that most stars have planets, and so our solar system is not special.

To say that Earth is special, having life that wasn't wiped out by a variety of dangers from space, seems presumptuous. However, there is no denying that Earth is in a good place within our galaxy, far enough from the center to lessen the danger, but close enough to the center to be in a star-forming region. Also, life occurred early enough from the inception of the solar system to allow complex life (even intelligent life) to evolve, which took billions of years, and our sun has enough fuel to continue burning for several billion years. If not special, then perhaps "just plain lucky" describes our planet.

References

Abbott, B. P. et al. (2017) "Multi-messenger Observations of a Binary Neutron Star Merger," *ApJ Lett.* 848, L13.

Alvarez, L. W., Alvarez, W., Asaro, F., and Michel, H. V. (1980) "Extraterrestrial cause for the cretaceous-tertiary extinction," *Science*, 280, pp. 1095–1108.

Arnett (1989) "Supernova 1987A," *Ann. Rev. Astron. & Astrophysics*, 27, pp. 629–700.

Collar (1996) "Quoted in "Did cancer kill the dinosaurs?", *New Scientist*, January 13.

Dolan, M. M. et al. (2017) "EVOLUTIONARY TRACKS FOR BETELGEUSE," *ApJ*, 819, p. 7.

Ellis, J., and Schramm, D. N. (1996) "Could a nearby supernova explosion have caused a mass extinction?" *Proc. Nat. Acad. Sci.*, 92, p. 235.

Gehrels, N. et al. (2003) "Ozone Depletion from Nearby Supernovae," *ApJ*, 585, pp. 1169–1176.

Knie, K. et al. (2004) "Fe-60 Anomaly in a Deep-Sea Manganese Crust and Implications for a Nearby Supernova Source," *Phys. Rev. Lett.*, 93, p. 171103.

Melott, A. et al. (2004) "Did a gamma-ray burst initiate the late Ordovician mass extinction?" *Int. J. Astrobiology*, 3, pp. 55–61.

Piran, T. and Jimenez, R. (2014) "On the role of GRBs on life extinction in the Universe," *Phys. Rev. Lett.*, 113, p. 231102.

Rampino, M. R. and Haggerty, B. M. (1996) "The "Shiva Hypothesis": Impacts, mass extinctions, and the galaxy," *Earth, Moon and Planets*, 72, pp. 441–460.

Wallner, A. et al. (2016) "Recent near-Earth supernovae probed by global deposition of interstellar radioactive Fe-60," *Nature*, 532, pp. 69–72.

Chapter 8
Life in the Balance

Life is like riding a bicycle. To keep your balance you must keep moving.
—Albert Einstein

Up to now, we have explored ways in which the universe has balance points. There are many tipping points where one could imagine a universe slightly different from our own, which would be inhospitable for life. Now it's time to bring this all together and look at what it all means. Previous chapters are about the question of "how" the universe is, but we still haven't gotten into the question of "why" it's this way. Of course, delving into "why" takes us away from the realm of hard science and into areas such as philosophy and religion. Can anyone really answer the question of *why* the universe is the way it is? Probably not. But we *can* observe how nature is so perfectly balanced as to allow life on Earth to exist.

Let's make a list of some tipping points we encountered in previous chapters:

1. A chance quantum fluctuation leads to inflation and the Big Bang.
2. A quirk of quantum mechanics, called the Pauli principle, balances the inward attraction of gravity, preventing all matter from collapsing into black holes.
3. Matter and antimatter are not in balance, leaving a tiny fraction of matter.
4. Neutrons tip the scale over protons, allowing hydrogen (and stars) to exist.
5. The nuclear force is "just right": if a bit weaker, the deuteron isn't bound; if stronger, then the di-proton is bound. Either way, the universe is very different.
6. Carbon is essential for life, but only exists due to a chance balance point between the mass of three helium atoms and the Hoyle resonance in carbon.
7. Heavier atoms, such as gold and uranium, likely come from the merging of neutrons stars, which only exist due to the balance of nuclear and gravitational forces.

Nature's Balancing Act. Ken Hicks, Oxford University Press. © Oxford University Press (2025).
DOI: 10.1093/9780197771471.003.0008

8. Once Earth formed, a tipping point for life occurred when simple organic compounds combined to form the first reproducing molecule—likely a ribozyme.
9. Once life on Earth started, there are numerous ways it could have been snuffed out, such as supernovae, gamma-ray bursts, asteroid strikes, or pandemics.
10. While life on Earth survived long enough for humans to evolve, we must still avoid man-made extinction events, such as nuclear war and climate change.

Each of these balance points must tip into the direction that favors us. Otherwise, life on Earth either doesn't happen or doesn't continue.

First, and perhaps foremost, is the question of why our universe exists at all. It was shown in Chapter 1 that the laws of physics are consistent with the universe starting as an unusually large quantum fluctuation, which gets amplified through the process of inflation to ignite the Big Bang. It is truly remarkable that the universe could appear out of "nothing," but the theory of inflation provides a plausible explanation. To understand this properly, we need a theory of quantum gravity, which would unify the mathematics of quantum physics and Einstein's theory of general relativity. But using the existing equations, we know that quantum fluctuations happen (based on many experiments that have verified quantum theory), and we believe that Einstein's theory is correct (again, based on many experimental tests and astronomical observations). Together, these two theories make it plausible that the whole universe started as a Big Bang. The recent discovery of gravitational waves demonstrates how gravitational mass can be converted into stretched spacetime, which is the core idea behind inflation.

Assuming the universe started from a quantum fluctuation, one can ask, "What is the probability of an unusually large quantum fluctuation to occur?" It's difficult to answer this because we don't have any experimental evidence or theoretical guidance for the properties of the particle that caused the quantum fluctuation. What we *do* know is that our universe has lasted for about fourteen billion years. So, it appears that the probability of this quantum fluctuation is likely to be less than once in the lifetime of the universe. But how would we know whether it hasn't already happened again? Perhaps the radiation from a newer Big Bang hasn't yet reached us, and we're all doomed in the near future. Of course, we wouldn't be here to think about it if *something* hadn't disturbed Nature's balance in whatever emptiness existed before the Big Bang!

Given that matter dominates the universe, we can ask why it doesn't all collapse into a bunch of black holes. In Chapter 2, the Pauli principle explains how matter supports itself, allowing the ground to be solid under our feet. In summary, the Pauli exclusion principle says that no two particles can overlap in the same region of space with the same quantum numbers. That's a fancy way of saying that electrons in atoms stay away from each other, due to quantum physics, and similarly protons and neutrons in the nucleus can't get closer for the same reason. So, for matter that comes together as planets and stars, the Pauli principle keeps gravity at bay. Under enormous pressure, like at the center of a supernova explosion, matter can collapse into a black hole. But without the Pauli principle, all matter would collapse under gravity, and our universe wouldn't exist. So, what causes this law of physics, named after theoretical physicist Wolfgang Pauli? We don't really know, but it's obviously important for life to exist.

The next big question is why the universe, having just started, didn't just vanish into nothingness. The laws of physics are, as shown in Chapter 3, symmetric for matter and antimatter. If we take our physical laws as currently understood, we can't explain the imbalance between matter and antimatter that we see today. Using today's understanding of cosmology, the imbalance is tiny, about one part in ten trillion. But without this imbalance, matter and antimatter would annihilate each other, and our universe wouldn't exist. This is another crucial tipping point. Again, we don't have a proper understanding of how our universe came to be dominated by matter. Physicists have found some instances of asymmetry in the laws of particles and their antiparticles (such as in the weak decays of neutral kaons), but this asymmetry is too small to explain the imbalance of matter and antimatter in the universe. Research into this area is a hot topic today. Perhaps future theoretical ideas, such as string theory or supersymmetry, may lead to an explanation.

At this point, you might be thinking that there's a lot more that we don't know about Nature than we do know. As the philosopher Socrates once said, "I know one thing: that I know nothing." Indeed, the more we learn about a subject like science, the more that fissures in our knowledge become evident. Another thing that we don't know about our universe is the stuff that we call dark matter. (It's not anything like regular matter.) What is dark energy (see Box 8.1)? The evidence that these things exist in our universe can fill several volumes, but how it fits into the standard laws of physics is still a big mystery. Putting this topic aside, as important as it is, let's continue with reviewing the tipping points for life as we know it to exist.

Box 8.1 Dark Energy

It is generally accepted by physicists that no one knows what dark energy is, but here I indulge in some speculation. Caveat emptor!

The idea that empty space has energy is disconcerting at first. However, if you look at other examples from Nature, the question of what "empty" means is not what you might expect. A finite volume of space is not empty in the mathematical sense. Instead, space always has energetic particles moving through it. Before speculating about dark energy, let's look at some known bound states.

In Chapter 4, we discussed the proton in terms of quarks and gluons and examined the fact that most of its "mass" is really energy (see Box 4.3). This changes our viewpoint of the proton (and the neutron) from being a dense ball of mass to one of a dense ball of energy. The energy is in the form of gluons, which are exchanged between the quarks, making the strong force field that binds it together. Of course, according to Einstein's equation $E=mc^2$, energy and mass are equivalent, so does it really matter? Maybe so if Nature repeats itself.

Consider now the atom, which has a dense nucleus (made of protons and neutrons) surrounded by a cloud of electrons. If the atom is measured by the size of its electron cloud, then it's mostly empty space. The space isn't truly empty because it contains the tiny electrons and some energy. The energy is in the form of photons, which are exchanged between the electrons and the nucleus. The energy density of photons is much less than that of gluons in the nucleus, but it's not zero. The energy held by the photons makes the electromagnetic field that binds the electrons to the nucleus.

Both above systems are small, where the effects of quantum mechanics are readily seen. What about larger systems? In the solar system, planets orbit the sun, and the effects of quantum mechanics are not seen. But does this mean that quantum mechanics doesn't apply at all? In a quantum view of gravity, the gravitational field is made from the exchange of gravitons, which are the quantum particles (packets of energy) similar to photons, except gravitons couple to mass instead of charge. Gravity binds the planets to the sun, and it's easy to think of that binding energy being held by the gravitons. The problem is that there is no quantum theory of gravity, and so for now the graviton is a hypothesis.

Nonetheless, we know that electromagnetic waves can be sent through empty space and carry energy. We also know that gravitational waves exist and that they also carry energy. Could the energy in gravitational waves, which are propagating through the universe since the Big Bang, account for part of the stuff we call dark energy? One thing is for certain: the space between stars is not empty of energetic waves, and this energy shouldn't be neglected.

Nuclear physics is a complex subject. Quite a lot is known due to careful experimental work over the past century. For example, physicists can measure the nuclear force between a proton and a neutron to an accuracy of better than 1% and can calculate this using mathematical models to a similar accuracy. Hence, we can do a thought experiment of what the universe might look like if the nuclear force is slightly changed in the theoretical models. What we find is that the simplest nucleus, made up of one proton and one neutron (called a deuteron), would not hold together if the nuclear force were decreased by about 5%. The basic fusion process in stars, where two protons merge to form a deuteron (plus a positron to conserve charge), would not be possible if the nuclear force were weaker by just 5%. Our sun would not shine, and life on Earth (if it could even start) would not flourish in that frozen environment.

Similarly, if the nuclear force were about 10–13% stronger, then two protons could bind together into a "diproton" nucleus (not possible with the real nuclear force). If this were the case, stars like our sun would burn out quickly—in about a million years—before life could form and evolve. This doesn't mean that other kinds of stars aren't possible (such as "D-burning" stars), but the bottom line is that our universe would be a very different place.

The fine-tuning required of the nuclear force is even closer in the stellar fusion process that produces carbon. Carbon is, of course, necessary for life as we know it. Carbon-12 has six protons and six neutrons and is made from the fusion of two nuclei, beryllium-8 and helium-4. The mass of these two nuclei, added together, is more than the mass of carbon-12, and the extra mass is converted to energy (recall $E=mc^2$). But this process wouldn't occur fast enough without the chance coincidence that carbon has a *resonance* (where carbon can absorb a quantum of energy) that matches the total mass of the fusing nuclei. Change the nuclear force by even a few percent either way (stronger or weaker) and the amount of carbon produced drops by orders of magnitude. Without the plentiful carbon produced in stars, how could life as we know it exist?

The nuclear force again becomes important to explain how heavy elements are produced. Without uranium and thorium, the Earth's core would be cold, resulting in the loss of Earth's magnetic field, which shields us from cosmic rays. There has been a revolution in the past decade in our understanding of how most heavy elements are produced. Currently, these elements are thought to be formed when two neutron stars merge together. Such mergers have now been observed, due to advances in the detection of gravitational waves at facilities such as LIGO. The existence of neutron stars is a delicate balance between the nuclear force, gravity, and the Pauli principle. That such objects exist in our universe is a prime example of the balance of Nature. Without Earth's magnetic field, life on land would likely not be possible due to the harsh radiation from cosmic rays.

Turning away from the nuclear force, there are other mysteries of how life (as we know it) started and how life has survived. For all the research that's been done in the fields of biology and chemistry, we still don't know how a batch of chemical building blocks spontaneously came together to form the first self-sustaining molecule that could replicate itself. The fact that all life as we know it uses only right-handed sugars and left-handed amino acids, whereas these chemical building blocks are found in equal handedness from laboratory experiments, suggests that life started only once. While there is currently no way to prove this hypothesis, the fact that all chemical building blocks used by living organisms have the same chirality remains a mystery.

Perhaps the biggest mystery of all is that life continues to survive on Earth, over billions of years. The universe is full of odd phenomena that could potentially wipe out all life. A close supernova explosion or a close gamma-ray burst that is directed toward Earth are examples of astronomical objects that have the potential to extinguish life. Another danger to human life, now well established, is a large asteroid collision. Recent astronomy observations have scanned the sky to establish (with high confidence) that such a collision is not going to happen this century. Of course, this hasn't stopped Hollywood from portraying this doomsday scenario. To say that Earth is lucky is perhaps an understatement, considering the many dangers that could have extinguished all life.

8.1 What Does It All Mean?

They say that beauty is in the eye of the beholder. The explanation for why there are so many balance points in Nature depends on your point of view. Those balance points could have tipped the other way, leaving a universe that either never existed or a universe that is inhospitable to life. Whatever explanation I write here may or may not agree with your perspective. Please indulge me as I speculate.

If you believe that the universe formed according to a deeper (unknown) mathematical theory, then you might feel that none of the balance points are there by chance. In that case, the universe could only have occurred in one way, due to physical laws that are immutable and yet to be discovered. There is some satisfaction to that point of view, as it suggests life would not depend on a dramatic series of coincidences to exist. It's also impossible to disprove that a deeper theory isn't out there. With this philosophy, you can't be proven wrong. Of course, belief in a deeper theory can also be unsatisfying, in that it professes ignorance of our current knowledge of physical law. Some people have said that all physical laws of any practical importance are already known. So what if we don't understand dark matter or dark energy! Whether you agree with that or not, the idea that there may be a deeper, all-encompassing theory of the universe has driven many scientists to search harder for an explanation, developing ideas such as string theory and other hypothetical models for the underlying structure of matter.

Another possible viewpoint is that there are many universes that can be created with different values for the physical constants (see below). For example, there could be another universe (possibly created from a different quantum fluctuation) that has a different balance of matter and antimatter or a different nuclear force. If space-time is really infinite, then there could be an infinite number of multiverses (see Box 8.2), and at least one of those will have the characteristics suitable for life. This belief in the multiverse can be quite satisfying in that life *does* depend on a series of coincidences, and we just happen to live in the only universe that is hospitable for life as we know it. The problem, of course, is that this philosophy gives a hit-or-miss quality to the meaning of life, which seems rather too vague for me. Again, one cannot prove that the multiverse concept is wrong. This idea has gained a fair following in recent years and is even appearing in popular movies.

Box 8.2 Is Infinity Real?

Mathematicians work in an imaginary world where everything is perfectly defined. For example, you can have a perfectly smooth sphere, even though no such object exists in Nature. In the real world, objects are made up of atoms, and when examined with enough magnification, they appear rough. Similarly, mathematicians can work with objects that are infinitely small (a point), but in real life, quantum mechanics say that subatomic particles have wave properties (i.e., are spread out). The bottom line is that the real world is not the same as the perfect imaginary world of mathematics.

continued

Box 8.2 *continued*

What about infinity? This is a mathematical concept that may not exist in Nature. For example, you might say that there are an infinite number of particles in the universe, but our universe has a finite size (corresponding to fourteen billion years of expansion) and the number of particles in it is also finite. Similarly, the idea of something being infinitely small (such as a black hole being a singularity—a point) makes no sense in a theory of quantum gravity (Greene, 2010).

So what does it mean when people say that the multiverse is infinite? The idea of an infinite space filled with an infinite number of "bubble" universes is one that boggles the mind. Of course, there is no proof that the multiverse exists, so this is an imaginary concept. Like the mathematician's world, the infinite multiverse is a creation of our imagination.

The mathematician Cantor explored the concept of infinity and found that there are countable infinities (like the integers) and uncountable infinities (like the real numbers). So, even the mathematicians wrestle with this concept. They are very careful in handling infinity in mathematical proofs, and perhaps scientists should be more careful when suggesting that Nature is infinite.

An infinite number of parallel universes in a multiverse is convenient if you want to argue that our universe is just one out of many possibilities. Then our universe, with conditions that are just right for life, had to happen somewhere in the multiverse, and so there is no need to explain further. This avoids the need for religious explanations or the possibility that a Grand Unified Theory exists. However, it invokes the concept of infinity.

Personally, I've never liked the idea of an infinite multiverse, but I may be in the minority. Perhaps there are just a very large number of parallel universes. But until science can provide evidence of even one other universe outside of our own, I'll view the multiverse as purely hypothetical.

If you are a religious person, you might want to believe that an all-powerful God is responsible for designing the universe (and the Earth) to be just right for human life. Obviously, this standpoint has a long history and has been adopted by billions of people over the past millennia. There is clearly some satisfaction that comes from the belief that a superior being is the reason for creation, as seen by the success of religious organizations over time. Like the other viewpoints above, one cannot prove it wrong (i.e., one cannot prove that God does not exist).

One disadvantage of all three viewpoints above is that none of them provide a prediction (or hypothesis) that can be tested by scientific observation. In the latter case, one cannot know the mind of God, so anything is possible. For the case of a multiverse, we cannot make observations outside of our own universe. So, there seems no way to test that hypothesis. In the first case, an underlying theory could potentially be developed to where it can make observable predictions, but science has a long way to go to make such progress. For the scientifically minded person, the idea of an underlying theory to explain all of the physical balance points of our universe is appealing.

Are there other possible viewpoints? Certainly, and I invite you to speculate on your own reasons for why there appears to be many balance points that all need to tip in the right way for human life to exist. These balance points span the disciplines of physics, astronomy, chemistry, and biology. One could add sociology as well, since we have managed so far to avoid nuclear war or other situations that could cause our own extinction, but that topic is beyond the scope of this book. If the physical universe is not mysterious enough, human nature has its own unique unpredictability.

8.2 The Multiverse Concept

As the saying goes, history has a way of repeating itself. As technology improves, and knowledge grows, it seems that our worldview keeps getting bigger. If you think back to medieval times, the world appeared to consist of the Eurasia continent and nothing more. Maps were drawn in those times showing distorted shorelines and at the edge of the map: "here be dragons." As shipbuilding technology improved, people realized that the Earth was round, but the common worldview had just Earth and the heavens (at the time unreachable except via God) and nothing else. When technology for optic lenses came around, the telescope was turned to the sky, and other planets were seen. Copernicus showed that the Earth was not the center of the solar system (a very unpopular worldview at the time), and so humanity's worldview expanded.

In time, telescopes improved to show that the Milky Way was made up of stars, which appeared to astronomers much as the sun would look like from a great distance. The worldview expanded to include the galaxy, with billions of stars. As telescopes got bigger, it was possible to resolve images of "fuzzy stars," which were soon understood to be galaxies such as our own.

The worldview expanded again to encompass the universe, with billions of galaxies. Technology further improved so that planets around other stars could be observed, showing that most stars in those billions of galaxies had planetary systems, again enlarging the worldview.

What could be next in this progression? What could be larger than the observable universe? Unfortunately, it seems highly unlikely that we will ever see outside our universe. After all, light speed is finite, and all astronomical observations are consistent with seeing objects that are part of our universe and not beyond. However, human imagination is unbounded, and the next logical step to this progression of increasing worldview is to assume that our universe is just one "bubble" in an infinite extent of space that include other bubbles, each starting from a vacuum fluctuation followed by a Big Bang. This multitude of universes, called a **multiverse**, is where current thinking in cosmology is headed.

Would each universe in the multiverse have the same laws as our own? This would be true if there is a true Grand Unified Theory (not yet known) that ties together all the unknowns, such as the strength of the nuclear force with the other forces and regulates the mass ratios of particles (and other constants) of the Standard Model. But does such a theory exist? The best candidate so far is **string theory** (Greene, 1999), which appears to unify gravity with the other known forces. String theory exists in a world with eleven dimensions, of which some of those dimensions are hidden from us.

String theory is fascinating, and entire books are written about it (for a very brief summary, see Box 8.3). One of the interesting aspects is that there are many (zillions) of solutions to the string theory equations, of which our universe might be one solution (Greene, 2010). Other solutions are possible, and some of those might have different constants of the Standard Model (such as the mass ratio of the up and down quarks). If those constants change much, a "bubble" universe might not be suitable for life. To understand this better, consider an egg carton, with twelve cups for eggs, but extend the egg carton in all directions to have billions of cups. Now throw a ping-pong ball into it, and it will find a random cup. That cup represents one solution to the string theory equations. But if thrown slightly differently, it might have landed in a different cup. If each bubble universe "lands" in a particular "cup" (or solution to the string theory equations), then our universe is different from the others in the multiverse. This would suggest that we happen to live in the one universe where the constants of nature are just right. It comes down to a random process, where no grand "designer" is necessary to have a universe where conditions are fine-tuned for life (as we know it) to exist.

Box 8.3 String Theory

The appeal of string theory is that it can unify the gravitational force with the other forces of Nature. Einstein's theory of general relativity has three dimensions of space and one dimension for time, giving four-dimensional space-time. It seemed natural, at least to mathematician Theodor Kaluza, to try adding another space dimension to Einstein's equations, and remarkably the new theory could describe both gravitation and electromagnetism (via the Maxwell equations). This idea of adding extra special dimensions was explored further and eventually led to modern string theory, which has nine or ten space dimensions, along with time. This version of string theory unifies all four forces of nature, but the "theory" is really a hypothesis because the solutions to the equations are very difficult and string theory has not yet yielded calculations that can be tested by experiments.

If string theory is correct, then why don't we sense the extra dimensions? Physicist Oscar Klein first suggested, in response to Kaluza's five-dimensional theory, that the extra dimension might be "curled up" at very small distances. To understand this, think of a line, which is infinitely thin, and replace it with a tube (like a garden hose). The tube has a cylindrical surface, so a one-dimensional line becomes a two-dimensional tube. Now shrink down the tube's radius to subatomic dimensions and you wouldn't see the extra dimension. In other words, that tube looks the same as an infinitely thin string.

Physicist Ed Witten, one of the developers of modern string theory, explains it this way: "I might say that the idea of extra dimensions might sound a little bit strange to anyone who hasn't studied physics. Anyone who has gone into physics professionally will know that there are many things that are a lot stranger than extra dimensions. General relativity is strange, quantum mechanics is strange, antimatter is strange. All these things are strange but true. Compared to a lot of things that have come true in physics in the past, extra dimensions are not such a radical departure" (Davies, 1988).

In addition to the appeal of unifying the known forces of nature, string theory has additional features that are intriguing. In this theory, particles can be considered as loops of string rather than infinitesimal points. Presumably, this loop exists in the higher dimensions that are curled up. So, instead of a point moving along a line, we have a loop on the surface of a tube moving along the length of the tube (as a crude example). Different particles are then just different vibrational resonances of the loop. This might explain why quarks and leptons each come in three families and how they are related to each other. It could potentially explain the rich structure we see in the Standard Model of particle physics.

continued

Box 8.3 *continued*

The reality is that modern string theory is still in the early stages of development and that much more theoretical work is necessary before any testable predictions can be made from its equations. String theory holds great promise but is enormously complex.

Whether string theory is correct, or if a Grand Unified Theory exists, is pure speculation. However, as technology improves and our ability to probe questions about the Standard Model get better, it may well become possible to determine whether the multiverse idea is reasonable or not. For now, the multiverse is one possible answer to the question of why our universe is particularly suited (with all of the coincidences described above) for life to exist. In other words, if all possible conditions exist somewhere in the multiverse, then one of those must be the conditions of our universe.

8.3 Are We Alone?

The question of whether we are alone in the universe has been around for a long time. One of the first serious attempts to answer this was attempted by Frank Drake in the early 1960s, when he held a conference on this topic at the National Radio Astronomy Observatory in Green Bank, West Virginia. This was an informal meeting, attended by just ten people, where Drake first presented an equation that combined estimates of various items that together give a probability that intelligent life exists elsewhere in our galaxy (the Milky Way). These estimates will be discussed below. In principle, if the estimates are good enough, then **Drake's equation** could provide a reasonable guess about whether alien civilizations exist in our galaxy. The problem is that some of the estimates are no more than wild guesses; so in the end, predictions from Drake's equation need to be taken with a grain of salt.

The first term in Drake's equation is the rate of **star formation** in our galaxy. This is something that astronomers can estimate with a good degree of accuracy. From studies of what astronomers call the initial mass function (IMF) of stars, the average star has about half the mass of our sun. Based on this, astronomers estimate that two to three stars per year are formed in our galaxy. There is some uncertainty here (about 50%), but compared with other estimates described below, this is not too bad.

The next two term in Drake's equation is the number of **habitable planets** per star. Again, not much was known about this back in the 1960s, but space telescopes and advanced technology have allowed astronomers to observe the effects of planets orbiting other stars. Again, there is some uncertainty, but at the time of writing, scientists estimate that about half of all sun-like stars have a rocky planet like Earth. This is based on more than 4,000 so-called exoplanets that have been found orbiting other stars by NASA's Kepler space telescope. Our galaxy has about a hundred billion stars, but not all are sun-like. Putting the numbers together, the estimate gives about two billion habitable planets scattered throughout our galaxy.

To give this some context, one of the closest stars to our sun is called Alpha Centauri, about four light-years distant (meaning it takes light, traveling at a speed of over 186,000 miles per second, four years to reach it). Actually, Alpha Centauri is a triple-star system, with a tight binary system (Alpha Centauri A and B) and a third star orbiting those two at a much larger radius (called Proxima Centauri). The latter star is known to have an Earth-like planet, discovered in 2016, orbiting it with a surface temperature that could be right for liquid water to exist. Although astronomers are not yet sure about water being stable on that planet, the mere fact that such a planet exists on the nearest star to Earth gives a sense of how prevalent Earth-like planets are in our galaxy.

For the next terms in Drake's equation, the estimates fluctuate wildly. What is the probability that life will form on an Earth-like planet? If life forms, what is the probability that it will continue to evolve, eventually into an "intelligent" space-going alien civilization? (I put intelligent in quotes, since this term needs to be defined properly, and here we'll assume that a life form that can travel into space is called intelligent.) And finally, if **intelligent life** evolves, how long will it last before going extinct? The answers to all of these questions are just guesses, since we have only one example of a planet where life started (Earth), which may or may not be an incredibly lucky example.

Suppose you're optimistic. Assume that half of all planets that are Earth-like will develop some form of life, and most of those will evade extinction events and evolve into intelligent life. Furthermore, let's guess that an intelligent alien civilization will last for 10,000 years before causing its own extinction. Putting the numbers together into Drake's equation suggests (an optimistic guess) that a few thousand alien civilizations exist currently in our galaxy, under these assumptions.

The problem with Drake's equation is that the uncertainty in the estimates is big. For example, if I assume that only one out of 10,000 Earth-like planets will develop life, then Drake's equation suggests (using the other estimates above) that less than one intelligent civilization currently exists in our galaxy. Similarly, if you assume that an alien civilization is likely to last for a million years rather than just 10,000 years, then one gets millions of alien civilizations in our galaxy. With a range of between one and a million for the number of intelligent civilizations in our galaxy, the game of using Drake's equation to answer the question "Are we alone?" becomes almost meaningless.

Of course, the Milky Way galaxy is only one out of an estimated hundred billion galaxies in our universe. That estimate is based on observations of the Hubble Space Telescope (HST), from a study called the Deep Field, where HST was able to look at very faint galaxies that are very far away. Hence, the probability that there is an alien civilization somewhere else in the universe is likely to be pretty high.

However, I would guess that we will never be able to communicate with a civilization in another galaxy because the distances are so great. Science fiction movies like to portray humans as zooming through the universe at speeds greater than light-speed, but that's just fiction. Unless there is a revolution in physics, which overturns Einstein's theory of relativity, it seems unlikely that faster-than-light travel will ever be possible.

8.4 Parting Words

From accounts in the media, you might have gotten the impression that scientists have already figured out all the answers to the beginning of the universe and the beginnings of life on Earth. Nothing is further from the truth! I wrote this book partly because I wanted to provide an account of the huge questions that still remain in the quest to understand why the universe is perfectly suited for life.

While it's tempting to say that there is an underlying theory of the universe (or to turn to other explanations, such as religion) to explain why "things are the way they are," such thinking is pure speculation and should be left for dreamers. Scientists continue the hard work of plugging along, step by step, to learn more about each corner of Nature where the unknown lurks. While it might seem like a hopeless journey, those prone to exploration know that every long journey starts by taking the first step. So it is with scientists, peering into cracks of the Standard Model of particle physics and other unexplained areas of physics, chemistry, and biology.

In a paper written in 1960, physicist (and Nobel laureate) Eugene Wigner wrote about the "unreasonable effectiveness of mathematics in the natural sciences." While some people might view mathematics as a creation of the human mind, apparently it is more than that. Mathematics is the language of Nature, as any physicist can tell you. So, to learn more about Nature, you need to speak her language. Words can only go so far in describing the laws of physics, which are embodied in mathematical equations. These equations are used to predict how particles will behave in situations that we cannot produce in the laboratory, such as in strong gravity near a black hole. In fact, black holes were predicted by Einstein's equations long before they were observed in telescopes.

Even though few equations were used in this book, hopefully you have come away with a sense of how much we know (and how much remains to be known) about the laws of Nature. We now know that the universe appears to be fickle and, if tipped the other way at various balance points, could have shunned life on Earth. My viewpoint is that we are lucky to be here, and we should work to save our planet while we still can.

References

Davies, P. C. W. and Brown, J. (1988), *Superstrings: A Theory of Everything?*. Cambridge: Cambridge University Press.

Greene, B. (1999), *The Elegant Universe*. New York: W.W. Norton.

Epilogue

Life appears to be a miracle of sorts. Every tree you see, every fish in the ocean, and every human being on Earth is here due to the most unlikely of circumstances. Whether human life will continue here on Earth is, to me, a serious question. Maybe the picture on the front cover of two galaxies merging (to form what looks like a question mark) is trying to tell us something.

The future of our species is at stake. It depends on you and your actions as well as those of countless others around you. Our planet is warming at an alarming rate because of the greenhouse gasses that are constantly being pumped into the atmosphere by human activity. Some of this is due to people's unawareness of the dangers posed by their combined actions, and some is due to people who don't believe in (or don't want to believe in) the science behind this issue. Some people may even know and understand the science but are thwarting it for their own personal gain (at the expense of the future population). Whatever the reasons, the facts are undeniable: the amount of carbon dioxide and methane in the air is increasing exponentially, and this has led to warmer ocean temperatures and the melting of ice fields.

There have always been people who warned that the end was near. When the atomic bomb was dropped, there were claims that world war would end human civilization. When pollution threatened to destroy the environment, there were stories of a future Earth that was too poisoned to sustain life. When artificial intelligence started to become a reality, movies were made showing how it would take over and possibly destroy all human life. None of these scenarios has come true (at least not yet). Somehow, humanity has found a way to avoid these doomsday scenarios. Will the same be true for the current crisis of our warming planet?

It seems a terrible waste if the universe formed with just the right conditions for life to exist, only for a self-aware species, once evolved, to find numerous ways to bring about its own extinction. Maybe that's why there is no evidence at present for intelligent life elsewhere in our galaxy. Perhaps any self-aware species would inevitably succumb to the fate of extinguishing itself. If so, then the Big Bang and everything that follows has all the earmarks of a Shakespearean tragedy.

Maybe you are a more optimistic person and believe that the human race will once again find a way to overcome the problem of climate change. Indeed, young people today seem to be keenly aware of the need to act. Both industry and government have made tremendous strides forward in the technology of renewable energy. Electric generation from wind and solar power now rivals the cost of burning fossil fuels. Nuclear power generation has made incredible progress in safety, making the power plants built in the 1960s look like the equivalent of a Ford Model T car compared with today's models.

Whether you are a pessimist or optimist about the future of humanity, I hope this book has made you think. Life on Earth is the only life that we know of in a universe that has just the right characteristics to make human life possible. What happens next is up to you.

Index

2 04